A Decade of
Reform

A Decade of Reform

Science & Technology Policy in China

International Development Research Centre, Canada

State Science and Technology Commission, People's Republic of China

INTERNATIONAL DEVELOPMENT RESEARCH CENTRE
Ottawa · Cairo · Dakar · Johannesburg · Montevideo · Nairobi · New Delhi · Singapore

Published by the International Development Research Centre
PO Box 8500, Ottawa, ON, Canada K1G 3H9

© International Development Research Centre / State Science and Technology Commission 1997

Canadian Cataloguing in Publication Data

Main entry under title :

A decade of reform : science and technology policy in China

Includes bibliographical references.
ISBN 0-88936-815-5

1. Science and state — China.
2. Technology and state — China.
3. Technology transfer — China.
4. Technological innovations — China.
I. International Development Research Centre (Canada)
II. Title: Science and technology policy in China.

Q127.C5D42 1997   338.951   C97-980110-9

A microfiche edition is available.

All rights reserved. No part of this publication may be reproduced, stored in a retrieval system, or transmitted, in any form or by any means, electronic, mechanical, photocopying, or otherwise, without the prior permission of the International Development Research Centre and the State Science and Technology Commission.

IDRC Books endeavours to produce environmentally friendly publications. All paper used is recycled as well as recycleable. All inks and coatings are vegetable-based products.

CONTENTS

Foreword ix
Letter of Transmittal xi
Introduction xiii

PART I. MAIN REPORT OF THE MISSION

Chapter 1. General Findings 3
S&T reforms 3
The problems of implementation 5
The overall process of reform in China 5
International factors impinging on China's reform process 6
The lack of clarity — even ambiguity — in some policy pronouncements 7

Chapter 2. S&T Policy Issues Currently Debated in China 9
The problem of duplication and lack of coordination 9
Issues related to industrial technology 10
Issues related to basic research 16
Issues related to resource use 22
Issues related to the financing of S&T activities 29

Chapter 3. Emerging Issues 31
Policy advice 31
Need for policy integration 31
Setting research priorities 32

A national system of innovation..34
Models of R&D and innovation..35
Role of national key laboratories, ERCs, and national research centres................36
Human-resource issues...39
Some benefits and costs of the S&T reforms for research institutions................40
Some final observations..41

PART II. DETAILED OBSERVATIONS OF THE MISSION

Chapter 4. General Observations on the Reform Process................45
Evolution of the reform process...45
Role of ministries in S&T decision-making....................................52
Central–local relations..54

Chapter 5. Innovation and a National System of Innovation...............57
Concept of a national system of innovation and its use as a policy framework..........58
Stakeholders in China's NSI..58
Roles of stakeholders in the functions of China's NSI...........................63

Chapter 6. Policies for Basic Research in China........................71
Current policy in China..71
Observations from the Mission..74
Some reflections on international experience and practice.......................78

Chapter 7. The High-technology Sector...............................83
Background to the reform process...83
Role of CAS in promoting high technology....................................86
The Torch Program and high-technology development zones.....................88
NTEs and TVEs...89
The 863 Program...94

Chapter 8. State-owned Enterprises..................................97
The SOE sector..97
S&T reforms and the SOE...102
Issues for the future..107

Chapter 9. Agricultural Research and Rural Development...............109
Policy agenda for agriculture..109
Progress in implementing S&T reforms in agricultural R&D......................110
Factors constraining the implementation of the S&T policy reforms................115
Research funding..120

Chapter 10. Environmental and Social Development 123
S&T reform policies and environmental development 123
Progress and difficulties in meeting S&T reform goals 124
Deepening the S&T reform process for the environmental sector 131

PART III. ACCOUNT OF FINAL MEETINGS IN BEIJING, SHENYANG, XI'AN, AND SHANGHAI

Chapter 11. Review of China's S&T Policies 135
Discussions in Beijing 135
Discussions in Shenyang, Xi'an, and Shanghai 152
Postscript 153

Appendix 1. Team of Experts 155
Appendix 2. Acronyms and Abbreviations 157
Bibliography 159

FOREWORD

One of the constant themes of our time is change. With science and technology (S&T) as the main driving force, our world is changing fast. The way to adapt ourselves to the changes around us is to reform and innovate. As a result of the reform in national S&T management systems over the past decade — part of the overall reform in China — our work force in the field of S&T is in a better shape and position to serve the needs of the rapid socioeconomic development of the country.

Ever since the beginning of our reform, we have been open to new ideas, both from home and from abroad. However, this is the first time we have invited an international team to give a comprehensive review of the reforms that have been going on for more than 10 years. During this endeavour, we organized many interviews and discussion sessions between the members of the international team and the stakeholders of the reform. This process served as a forum for the exchange of ideas. The process of discussion itself was as important as the conclusions. In fact, several hundred people from government agencies, research organizations, universities, and industry participated in the discussions.

The original goals of this international review were to get views on our reform from a different angle, to draw on international experiences, and to search for effective policy instruments and reform directions as input to our policy measures for the years ahead. Through my personal experience of participating in the review at every stage, I can say that we have achieved the goals we set at the start. The review has been so successful that even before this final report is officially published, some of the ideas and concepts that emerged during the process have already been integrated into policy formulation.

Here I would like to thank Dr Keith Bezanson, President of the International Development Research Centre (IDRC), and Dr Song Jian, State Councillor and Chair of the State Science and Technology Commission (SSTC), for initiating this joint project between IDRC and SSTC. I would also like to express my sincere gratitude to Prof. Geoffrey Oldham, Science Adviser to the President of IDRC, and to every member of the International Team for the hard work, artistic provocation of discussion, and expertise in different aspects of S&T policy that they put into this investigation.

The success of this project owes much to the active support and participation of many organizations in the S&T sector and of other institutions, such as the Chinese Academy of Sciences, the National Natural Science Foundation of China, the China Association for Science and Technology, the Chinese Academy of Engineering, the local Science and Technology Commissions of Beijing and Shanghai municipalities and Liaoning, Shaanxi, and Guangdong provinces, Qinghua University, Anshan Iron and Steel Works Corporation, Beijing Yanshan Petrochemical Corporation, Shenyang Blowers Factory, and others that I will not list here one by one. I would like to thank all those people who helped to organize this project and those who participated in the discussions.

The S&T management-system reform is an ongoing process. I am grateful to the International Team members for suggesting, toward the end of this project, a number of follow-up options to improve the S&T system in China. Their offers have been warmly received by SSTC. The two topics we plan to concentrate our efforts on in the immediate future are a national system of innovation and a strategy for international collaboration on S&T. I welcome the continued support and participation of IDRC and the wider international S&T community as we strive to implement two long-term national strategies, namely, the strategy of revitalizing China through science and education and the strategy of sustainable development of China.

Zhu Li-lan
Member of State Council Leading Group on Science and Technology
Executive Vice Minister
State Science and Technology Commission
People's Republic of China

LETTER OF TRANSMITTAL

Zhu Li-lan
Executive Vice Minister
State Science and Technology Commission
People's Republic of China

Dear Madam Zhu:
I have much pleasure in enclosing our S&T Review Mission report, *A Decade of Reform: Science and Technology Policy in China*. Our observations and impressions are based on the written material translated into English by SSTC staff and mostly derived from the 1995 "White Paper on Science and Technology Policy," together with the 3-week visit we made in November 1995. During that visit we met with a wide range of science and technology (S&T) policymakers, those concerned with policy implementation, and those effected by S&T reform, in the provinces of Shaanxi, Liaoning, and Guangdong and in the cities of Beijing and Shanghai. These personal observations and impressions have been supplemented by some additional impressions gained by reading recent secondary reviews.

In reaching our conclusions we followed the process adopted by many similar reviews carried out in the member countries of the Organisation for Economic Co-operation and Development. We have relied almost exclusively on what has been written by Chinese authors and what we were told by Chinese individuals. We have not made our own detailed evaluations of institutions or programs. Our report is a mirror reflecting back what we read or were told. The filters through which information passed on its way to the mirror, however, were our own experiences and knowledge of our own societies. This influenced the questions we asked and our interpretation of the replies. We hope that our mirror is slightly concave so that our views are focused!

We have included a brief description of Chinese S&T institutions and also of the history of the reforms. We felt it was important that we state what our understanding is of these issues. If we are wrong in that understanding, our impressions may be distorted. We have also focused our report on the impact of the reforms with respect to the five themes you had asked us to pay particular attention to: basic research; the high-technology sector; traditional state-owned enterprises; agricultural research and rural development; and environmental and social development.

As you know, it was never the intention of the S&T Review Mission to make formal recommendations to the SSTC. Our task was to raise issues and ask questions. We hope our report makes a modest contribution to the way in which you and your colleagues consider the evolution of the S&T policy reforms.

Finally, I would like to express our sincere gratitude to the many people throughout China who gave their time so generously to answer our many questions.

Yours sincerely,

Geoffrey Oldham
Head
IDRC / SSTC Science and Technology Review Mission to China

INTRODUCTION

The origins of this particular review of China's experience in the reform of its science and technology (S&T) system go back to a conversation between the Chair of the State Science and Technology Commission (SSTC), Song Jian, and the President of the International Development Research Centre (IDRC), Keith Bezanson, held in Beijing in 1994. They noted that 1995 would be both the 10th Anniversary of some of the major S&T reforms in China and the 15th Anniversary of cooperation between SSTC and IDRC in the financing and management of a program of research and development (R&D) in support of China's development. To mark the dual anniversary, they decided to jointly commission a review of the Chinese experience of S&T reform, particularly in the last decade, and agreed that the methodology best suited to the task would be one developed by the Organisation for Economic Co-operation and Development (OECD) in Paris for reviewing the S&T policies of industrialized countries.

As a result of a detailed agreement between SSTC and IDRC, the following terms of reference were set for the S&T Review Mission to China:

> Terms of Reference
> Components and Objectives of Evaluation
> The overall objective of this review is to assist Chinese policymakers as they assess the impact of their reform policies and decide on whether modifications to these policies should be made. The review will focus primarily on the science and technology policy reforms initially introduced by the SSTC in 1985.
> The review will consist of three phases:
> a) The gathering of background information for the project as discussed with Mr James Mullin, a consultant engaged for this purpose, in early July

1995. The project will be guided by the standard OECD country review formula;
b) The review visit of the international team towards the end of 1995; and
c) The final meetings.

After establishing these terms of reference, the two sides agreed on five substantive areas of concentration for the review and on the composition of a seven-person international team of experts (see Appendix 1) to carry out the review.

The specific areas for priority attention were basic research; the high-technology sector; state-owned enterprises (SOEs); agricultural research and rural development; and environmental and social development.

Method of work

The review was carried out in three steps. First, a preliminary visit to Beijing was made by one of the team, in July 1995, to reach agreement on the background documentation to be provided by SSTC to the Mission. It was agreed that SSTC would provide English-language documentation, including translations of key chapters of the recent "White Paper on Science and Technology Policy," together with translations of a series of policy studies from the National Research Centre for Science and Technology for Development (NRCSTD) and the Institute of Science Policy and Management of the Chinese Academy of Sciences. In addition, copies of the English-language edition of NRCSTD's journal *Forum on Science and Technology in China* were also made available.

Second, the Mission visited China during the period 9 November to 1 December 1995 for an extensive series of meetings with scientists, engineers, managers, and officials. These in-depth interviews formed the basis for this report, which was drafted by the Mission at the completion of its stay in China. The intention of the review was to look at the S&T reforms in China in the light of our understanding of how patterns of research and innovation are evolving in industrialized countries. It was not a task of the Mission to recommend particular policy options for China — that task is appropriately the responsibility of the Government of China and of the Chinese institutions engaged in S&T.

Third, a final meeting was held in Beijing in May 1995 to permit SSTC officials to discuss with the Mission its findings and impressions. A set of notes on the meeting appears as Chapter 11 of this report.

Some caveats

Readers of this report should bear in mind the highly selective nature of the evidence gathered during the work of the Mission. Given the brevity of the visits, there were many important areas of Chinese S&T activity entirely untouched by the Mission. For example, we were unable to touch on areas of science like medicine or forestry; we visited no "big-science" facilities; and we discussed, without visiting, a number of key new institutional forms, such as the key national laboratories and the engineering research centres that were created during the decade of reform.

We have also encountered a problem interpreting Chinese statistics, some of which (for example, those on S&T "achievements") are based on definitions with which we are unfamiliar.

Despite these caveats, we happily acknowledge the great openness of our Chinese hosts, who often went to great lengths to answer our many questions.

Structure of the report

The report is divided into three principal parts.

"Part I — Main Report of the Mission" represents our attempt to summarize the principal impressions we had of S&T reform in China. We begin with a short series of general impressions (Chapter 1), which provided the backdrop to our understanding of what has transpired in Chinese S&T over the last decade. We then attempt to summarize the principal debates on Chinese S&T policy that we encountered during the Mission (Chapter 2), adding, where possible, some remarks on relevant international experience or practice that we believe could be useful in the present context in China. Then, we raise some issues that we believe are not receiving the attention in China that, we would argue, they merit (Chapter 3). Finally, we add a few brief observations on current S&T policy in China as articulated in the Communist Party of China Central Committee and State Council Decision on Accelerating Scientific and Technological Progress (6 May 1995). In Part I we also highlight 10 key issues we believe to be particularly important at this time and worthy of further discussion.

"Part II — Detailed Observations of the Mission" begins with a lengthier discussion of our understanding of the S&T reform process in China (Chapter 4). We then introduce what we consider to be an important concept — the concept of a national system of innovation (Chapter 5) — which is helpful for conceptualizing the range of functions needed to fully attain the goals of China's S&T reforms. We then provide commentary on the five specific areas designated in our terms of reference: basic research

(Chapter 6), the high-technology sector (Chapter 7), SOEs (Chapter 8), agricultural research and rural development (Chapter 9), and environmental and social development (Chapter 10).

"Part III — Account of Final Meetings in Beijing, Shenyang, Xi'an, and Shanghai" contains notes on discussions held in May 1996.

Part I

Main Report of the Mission

Chapter 1

GENERAL FINDINGS

S&T reforms

Since at least 1978, China has encouraged experimentation in its science and technology (S&T) system as a means of arriving at reforms and has periodically summarized the main directions of reform in authoritative decisions of the State Council and of the Central Committee of the Communist Party of China (CPC), particularly the March 1985 Decision on the Reform of the Science and Technology Management System (Box 1) and the May 1995 Decision on Accelerating Scientific and Technological Progress (see Box 2). This has been a creative method of approaching a complex set of issues, and the main decisions have established a sensible overall framework for S&T policy for a modernizing economy. In some cases, such as in environmental policy, the decisions already taken in China are ahead of those taken in most other countries in the world.

The 1993 Decision on Issues Concerning the Establishment of a Socialist Market Economy Structure appears to have injected added impetus to S&T reforms in some parts of China and has given rise to another round of creative experimentation.

We have taken the March 1985 Decision as our main point of departure and have also considered the effects of other reform initiatives indicated in the time line shown in Figure 1.

Box 1

The Decision on the Reform of the Science and Technology Management System (March 1985)

I Modern science and technology constitute the most dynamic and decisive factors in the new productive forces .... We should reform China's science and technology management system resolutely and step by step in accordance with the strategic principle that our economic construction rely on science and technology and that our scientific and technological work must be oriented to economic construction.... Regarding the operating mechanism, it entails reforming the funding system, exploiting the technology market and overcoming the defects of relying on purely administrative means in science and technology management, with the state undertaking too much and exercising too rigid a control.

II Funding for research institutes should be reformed so as to practise classified management over the operating expenses which is suited to different types of scientific and technological activities. [This led to different state-funding responsibilities for different kinds of S&T activities and institutions.]

III We should promote the commercialization of technological achievements and exploit the technology market so as to suit the needs of the socialist commodity economy.

IV In restructuring the science and technology system, emphasis should be placed on encouraging partnership between research, educational and designing institutions on the one hand and production units on the other and on strengthening the enterprises' capability for technology absorption and development.

V The management system in agricultural science and technology should be reformed so as to serve the restructuring of the rural economy and facilitate its conversion to specialization, commercialization and modernization.

VI To ensure sustained progress in economic and scientific and technological development, it is necessary to deploy our scientific research forces rationally and in depth.

VII More decision-making power should be granted to research institutes, and macromanagement of scientific and technological work by government organs should be improved.

VIII Opening to the outside world and establishing contact with other countries is a basic and long-term policy in China's scientific and technological development.

IX Management of scientific and technological personnel should be reformed to create a situation favourable to the emergence of large numbers of talented people who can put their specialized knowledge to best use.

Source: SSTC (1986)

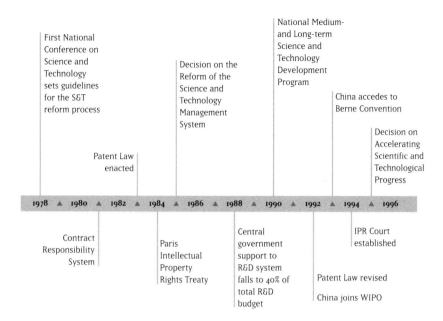

Figure 1. Time line of significant policy events in the reform of China's S&T system.

The problems of implementation

Although the Mission believes that the broad directions of reform set out in the State Council's decisions are appropriate to China's overall development, we have observed that there have been great variations in the level of implementation of reforms in different parts of the country and in different institutions. Not surprisingly, in the eyes of the Mission, there is a strong correlation between successful and innovative attempts at reform and the presence of strong entrepreneurial management. Equally, where management retains the mind-set of the command economy, reform has been slow.

The overall process of reform in China

The reform of China's S&T system is only one element of an ongoing and long-term process of reform of its national economy and social-welfare system. China is emerging from a command economy that has produced, contrary to its intentions, a huge and, to the outsider, confusing array of institutions in all sectors that are almost invariably overstaffed and often duplicate the work done by others. There is a need to continue the process of rationalization that began with the reduction of government support to many of these institutions. However, this process is substantially impeded by the significant overhead costs that institutions incur in providing housing, health care, education, and other services to their employees, both current and retired, and to their families. Key to tackling these important barriers to institutional reform will be the as-yet incomplete process of reform of the social-welfare system. Until a new system, geared to the demands of the emerging socialist market economy, is in place, many necessary structural changes, including changes in the S&T system, will not be possible.

International factors impinging on China's reform process

When China decided to develop its "open-door policy" in the late 1970s, one of the main reasons was to acquire access to foreign technology and foreign management methods. It was expected that these acquisitions would boost the Chinese economy and bring improved living conditions to the Chinese people. It was also expected that encouraging Chinese scientists to study and work abroad would lead to an inflow into China of foreign knowledge, technologies, and organizational and management practices. All of this has happened, and on a scale scarcely imaginable 15 years ago.

During these 15 years, not only has China changed, but so too has the rest of the world. These changes have led to increased globalization, not only of economies, but also of S&T. International collaboration in science is increasingly the way in which science is practiced, and the cost of doing research has meant that international collaboration is also necessary for sharing costs. Technological collaboration within business cooperative arrangements is also becoming more global. Multinational corporations frequently do their research in a first country, their design in a second, their development and production in a third, and their initial sales in a fourth. Unless a country has technological assets that enable it to be an attractive partner in this globalization, it will become increasingly isolated.

China clearly has many S&T assets; it has embraced its open-door policy; and it has paid attention to the encouragement of international collaboration. Nevertheless, the Mission formed the impression that there is not an explicit policy in China for international collaboration in S&T that fully embraces the implications of today's global realities in technological development. Given the subject's importance, China could consider convening a small international conference on national strategies for international collaboration in S&T. This would be a way of learning about other international experience on an issue of major significance for China.

The lack of clarity — even ambiguity — in some policy pronouncements

We discuss briefly in Chapter 6 the use of eight-character slogans to convey the essence of policies — a practice whose result is to allow considerable freedom for interpretation to those decision-makers who are willing to be entrepreneurial and innovative. One such slogan, *wenzhu yitou, fangkai yipian*, informally translated by the Mission as "anchor one end securely, let the other roam free," is just one example of a policy statement that is subject to many interpretations.

The Mission recognizes that these phrases are designed to express in a concise fashion the essence of the longer policy statements that they accompany. Additionally, these phrases can be seen as setting boundary conditions within which a good deal of experimentation can take place. This stimulation of local experiments has, in fact, been one of the strengths of the Chinese approach to reform.

Chapter 2

S&T POLICY ISSUES CURRENTLY DEBATED IN CHINA

The problem of duplication and lack of coordination

With China having literally thousands of S&T institutions, it is not surprising that there is duplication of effort throughout the system. The issue that now needs to be addressed is the duplication within a socialist market economy.

A competitive system among enterprises in the future will ensure that a variety of common problems are attacked from a variety of competing perspectives. This may lead to a superficial appearance of duplication, but the market, if allowed to operate, will weed out the unsuccessful.

However, a system of extensive duplication of effort in governmental institutions that have no tradition of cooperation with each other is an expensive luxury that no country can afford. The Mission has seen examples of extensive duplication of efforts among uncoordinated, separate institutions — all funded by one level of government or another — that coexist in close proximity.[1]

---

[1] In one province, we heard of efforts to coordinate some 42 independent institutes in a program of rural development with fairly specific interests.

Issues related to industrial technology

Many of the policy discussions we encountered during our visit to China related to the commercial application of research results. In the following sections we identify a number of current debates related to industrial technology and then attempt to highlight, where possible, policy issues that need continuing attention as part of the ongoing process of S&T reform.

Introduction of a "technology market" and of competitive sources for funding

An essential element in the move away from the command economy of the past was to create a market within which users of new technology would become important sponsors of its development. Government moved swiftly in the period between 1987 and 1992 to inject the research and development (R&D) institutions into this technology market by

- Substantially reducing the institutions' core-budget appropriations from government (in some cases, especially in the provinces, this reduction was to zero);

- Making government-contract funds, allocated by competition, an important vehicle for promoting technological development; and

- Providing incentives for enterprises to invest in R&D at an increased rate.

This market discipline has had the positive effect of allowing creative institutions to expand their incomes and activities and to have a substantially increased impact on the economy.

Not surprisingly, many enterprises were slow to respond, and this has created great difficulties for most R&D institutions. Many enterprises are unwilling to pay for technology that in the past they would have received without charge, or they seriously undervalue the technology that has been on offer from R&D institutes. (Similar experiences have been observed in industrialized countries.) This "market failure" has influenced a number of R&D institutions to transform themselves into enterprises to commercialize their own technologies.

The Mission found that entrepreneurial scientists and institutes were very much in favour of these reforms because they had given them the opportunity to increase the resources available for their work, although some of those who had been less successful in the competitive arena were less enthusiastic. In some provinces, there appeared to be a growing belief, even among successful provincial institutes, that only national institutes could win support from some of the programs offered by the State Science and Technology Commission (SSTC) (such as the 863 Program or the National Program for Tackling Key Technology Problems). In other regions of the country, we did not encounter this view. It would be worthwhile for SSTC to periodically review competitions for national funds to be assured that decisions are made on the technical merits of proposals and not simply on the organizational status of the group applying.

Chinese experimentation with the concept of a technology market may well lead to some distinctive features in what we refer to as China's national system of innovation (NSI) (see Chapter 3). One critical question in any market is that of minimizing the transaction costs of interactions between organizations (for example, enterprises and research institutes) that have to deal with each other. Can an appropriately designed policy environment facilitate low-cost interactions between enterprises and suppliers of new technologies? Some continuing attention to this issue could help create a positive climate for innovative activity.

Emergence of spin-off enterprises

A very common response of R&D institutions to the new market conditions has been to create spin-off enterprises in attempts to commercialize technologies that they have developed. A very large number of these enterprises have been created — the Chinese Academy of Sciences (CAS) informed the Mission that the Academy and its 123 institutes had created 900 such spin-offs! Their rate of success appears to be similar to that in industrialized countries — that is, about 1 or at most 2 in 10 are very successful, 2 or 3 can survive in the longer term without much expansion, and the rest fail. However, the present policy in China does not allow these failed enterprises to go out of business because that would adversely affect the social welfare of the employees involved. Vigorous debate on economic reform, as it touches on the problem of bankruptcy, is continuing: for example, researchers at the Economics Research Institute of the Chinese People's University in Beijing argued that "poorly operated enterprises should be allowed to go bankrupt so that inefficiently used funds can be freed to be injected into profitable enterprises" (*China Daily* 1995, p. 4).

Responsibility for continuing to subsidize unsuccessful spin-off enterprises appears to reside with the parent institute. The Mission believes that if China is to develop a successful socialist market economy in which new technology enterprises (NTEs) play an increasingly important role, it will be necessary to speedily resolve the question of a new social-welfare system and to permit the process of bankruptcy to occur.

Imported technology

We are aware that China has imported very substantial amounts of foreign technology in recent years, particularly through programs of technological revitalization of enterprises. In some enterprises we visited, there are vigorous programs in place to integrate those technologies into the fabric of the enterprise's production system and to build on this technical basis. We have heard of specific cases in which enterprises may spend as much as three times the purchase price of foreign technology on programs to master, adapt, and build on that technology. We are, however, not in a position to judge how widespread this practice is, particularly among state-owned enterprises (SOEs), but other evidence suggests that it is not common. For example, the Science and Technology Commission (STC) of the Chinese People's Political Consultative Conference contends (STC–CPPCC 1994, p. 48) that

> the funds for digestion, assimilation and innovation are seriously inadequate. In Japan and South Korea, the expenses for digestion and assimilation are much higher than that of introduction of technology, but in China, on the contrary, when 1.00 yuan is spent on introduction of technology, only 0.09 yuan is used for digestion and assimilation, and in Shanghai it is only 0.07 yuan.[2]

We believe that future reforms should look for means to encourage investments in mastering and building on imported technology.

Organizing for technological innovation within enterprises

Although the Mission heard much comment about the need to invest in R&D within enterprises, we heard little debate about how this R&D capacity should be organized. Also, as the comments of STC–CPPCC (reported above) underline, a few enterprises have recognized the need to invest resources to properly absorb and master imported technology, but many have not. In the opinion of the Mission, these two issues can be interrelated in important ways.

In industrialized countries, many large corporations have recognized that there are two types of technical change. One is radical technical change, which usually requires formal R&D laboratories, but the other is incremental technical change, which may sometimes involve a formal R&D organization but frequently is introduced by teams of specialized engineers who work close to the production units. Sometimes the production units themselves introduce these incremental changes. The specific way these incremental-technical-change units function varies from enterprise to enterprise and from industry to industry.

---

[2] In 1996, 8.07 yuan renminbi (CNY) = 1 United States dollar (USD).

Some industries and enterprises in China are aware of these issues. For example, the Chinese National Offshore Oil Corporation worked with a UK research group a few years ago to understand the process of innovation in UK offshore drilling companies. However, we formed an impression that, over the past 5 years or so, the majority of Chinese enterprises that have imported technology have done so without investing in the activities necessary to absorb and assimilate that technology. Chinese enterprises and policymakers may find it worthwhile to investigate the experiences of corporations in industrialized countries that have successfully assimilated imported technology and organized for both radical and incremental technical changes.

One point worth emphasis relates to enterprises' strategies to promote technical change. It is important for enterprise managers, operating in a global-market context, to understand the limitations to what governments can do for them and the crucial roles that they, themselves, will play in determining the technological success or failure of the enterprise for which they are ultimately responsible.

Protection of intellectual property

The Mission found that Chinese institutions are beginning to see the practical benefits of a system for the active protection of intellectual property. Some institutes have already had to prosecute employees for making unauthorized transfers — typically to township and village enterprises (TVEs) — and have found it difficult in the courts to get the legal system to understand the basis of their case. In other cases, spin-off enterprises have had to be careful about the ownership of the technology they take into international markets. As China moves progressively into global markets and moves to freeing trade domestically, it will need to ensure the adequacy of its system for upholding the property rights that its legal system confers.

An issue for China to consider is the adequacy of the present system of implementation of its intellectual property legislation. A recent World Bank (1995, p. 10) document argues that among the remaining challenges (of S&T reform) facing the government of China is intellectual property rights (IPR) enforcement. The document suggests that

> the key concern is how effectively intellectual property legislation is and will be enforced. The key tasks ahead involve raising awareness of intellectual property rights, among users and providers of technology, fostering the further professionalization of enforcement personnel, and ensuring that these rights are adequately pursued. A nationwide professional court system is needed, exemplified by the recently created Intellectual Property Courts dedicated to enforcing the IPR regime. Education and dissemination seem to be key to improved enforcement in the short term.

High-technology development zones and NTEs

China now has 52 high-technology development zones, spread across the country. Within these zones a variety of national and provincial incentives operate to encourage the development of NTEs. There are estimated to be 55 000 "approved" NTEs in these zones.

Some of these zones, in provincial cities, are still small, but others, such as the one in Pudong Development Area in Shanghai, are very large. Additionally, China has several other categories of zones within which enterprises are eligible to receive preferential treatment (for example, the Pudong Development Area consists of a finance and trade zone, an export-processing zone, a free-trade zone, and a high-technology zone, all of which seem to be competing to attract the same enterprises and have similar sets of incentives available).

According to a draft report from the Institute of Science Policy and Management (ISPM) of the CAS (Fang Xin n.d.), there are six important sets of issues relating to these new zones and the companies that operate in them:

- The ownership of the assets of the NTEs, most of which have been created by R&D institutes and other public bodies (in the absence of a fully developed share-holding system, the ownership of the assets of many NTEs can be in doubt);

- The modernization of the management of many NTEs;

- The precise delimitation of the roles of the management authorities of the new high-technology development zones (Do they constitute a new level of government? Are they regulatory agencies, determining the eligibility of individual enterprises claiming the incentives offered by the zones? Or are they support systems, designed to help NTEs through the early stages of enterprise growth and development?);

- The treatment, from an incentive viewpoint, of enterprises within a zone that expand their production in facilities outside the zone;

- The eligibility of NTEs to compete for further national support through a variety of funding channels; and

- The development of adequate systems of taxation administration, financial management, accounting and auditing, both for NTEs and for the zone-management authorities.

In one province, Guangdong, the Mission saw evidence of an apparent concentration, in the Pearl River delta, of high-technology parks and zones, some national and others provincial,[3] and of plans to create a much more

---

[3] According to provincial officials, the incentives in a national zone are "somewhat better" than those available in a provincial zone.

extensive high-technology belt. The subtleties in the distinctions between *park*, *zone*, and *belt* were never satisfactorily explained to the Mission.

Two points can be drawn from international experience that may be relevant to the future development of high-technology development zones. First, in industrialized countries, there is a growing appreciation of the effects of geographic clustering of complementary industrial competencies. In a world in which globalization is increasingly perceived as the driving force behind enterprise strategies and in which national boundaries seem to be less and less significant, it is also becoming clear that the competitive success of many enterprises is due to localized concentrations of skilled people and technologies. To the extent that China's high-technology development zones can produce focused concentrations of skills and technologies and can promote cooperation among the enterprises located within those zones, they may in fact be important contributors to the global competitiveness of those enterprises that have the managerial capacities to succeed.

Second (on a less positive note), as China moves to join the World Trade Organization (WTO), it should look at the WTO treatment of subsidies for R&D. The kinds of incentives China offers within specified geographic locations (and hence not open to enterprises located elsewhere) may be deemed as unacceptable to the WTO and, as such, could render products produced in the zones vulnerable to countervailing tariffs. It is by no means clear that this would in fact be the case, but Chinese authorities should be attentive to this possibility.

Transfer of military technology for civilian use

China has had a policy of encouraging the transfer of military technology for civilian use for 15 years. The recent "White Paper on Arms Control and Disarmament" (IOSC 1995) revealed just how extensive the transfer has been. More than 15 000 products for civilian use are produced by defence establishments. These include 60% of all motorcycles, 9% of all automobiles, and 24% of all mechanized coal-cutting equipment.

The Mission visited one military R&D institute (the Northwest Research Institute of Electronic Equipment) that is producing satellite antennae for the civilian market. This establishment had moved to Xi'an to take advantage of the preferential tax incentives offered by the high-technology development zone. The Institute markets its products both nationally and internationally. Profits from the civilian production are plowed back into the civilian work, and, in practice, part of its revenues help finance some of the huge social costs that the Institute bears.

The policy issue of interest to the Mission was whether the Chinese approach of transferring military technology for civilian use by means of defence-establishment manufacturing of civilian products is the best approach. In international experience, a different approach has been taken — military technology is transferred to civilian enterprises; thus, military R&D establishments continue as R&D establishments and rarely get involved in production. Civilian companies manufacture armaments or civilian products, or both. In industrialized countries, it is believed that civil enterprises, with management systems in tune with the marketplace, are better placed to manufacture products for that marketplace than defence industries that have usually operated on a cost-plus basis.

The Mission recognized that without changes to the social-welfare system, the Chinese approach is probably the most sensible one. If social reforms make it possible to humanely shed staff, then it would be worthwhile for China to study the foreign experience of technology transfer and the foreign experience of developing dual-use technologies, that is, technologies that can be exploited for both military and civilian uses.

Product design and quality

China is no longer shielded from competition, and its products and processes must be able to compete with overseas technology, both on the Chinese domestic market and in global markets. Programs to upgrade product design and quality and to improve market intelligence should be features of the next round of S&T reform. We believe that first steps in this direction are being taken within a World Bank Technology Development Project (World Bank 1995). Attention to industrial design and participation in international quality-assurance schemes, such as the International Standards Organization's ISO-9000 Program, should be increasingly seen as desirable options for enterprises participating in China's programs of technological revitalization and S&T reform.

Issues related to basic research

A second set of issues we encountered related to the future of basic research in China.

Some questions of definition and their policy implications

In industrialized countries, basic research has come to embody a whole range of activities:

- So-called curiosity-oriented research, often, but not always, carried out by individual scientists searching for general scientific understanding rather than seeking to contribute to the solution of some identified social or economic problem — Such research encompasses whole fields of science, such as cosmology and astronomy, and is also practiced at the boundaries of knowledge in other fields in which the mainstream is already linked to application. Such research in general shows little likelihood of contributing to economic development at any foreseeable time. This type of scientific activity is giving rise to informal international networks of individuals and groups whose collaboration is facilitated by inexpensive electronic communications. The activity also provides a useful opening into the global scientific community. In industrialized countries, curiosity-oriented research is widely seen as a good vehicle for training new generations of researchers because it provides ample opportunities for young scientists to learn research skills. Many academic programs still demand that candidates undertake an individual piece of original research; large numbers of graduate students are the hallmark of successful programs of this type.

— Strategic research, in which teams of researchers, frequently from a variety of disciplines, explore the frontiers of knowledge in broad areas of science that are believed likely to be of future economic or social importance — In this domain, the probability of some application is thought to be foreseeable in the medium term, although the specific route to application might be far from evident at this time. This form of research is becoming more important (but should never replace curiosity-driven research in its entirety) and is posing new problems for institutions of higher education, which have to devise new ways of evaluating the research of graduate students who work in teams rather than alone and who work in new transdisciplinary areas rather than within the confines of a traditional academic discipline.

— "Big science" of two kinds: science, such as high-energy physics, that requires ultraexpensive facilities; and the geographically extensive research needed to understand changing global environmental phenomena — Both types of big science are often affordable only by consortia of countries prepared to share the considerable costs involved.

We believe that policy options for the support of basic research in China would become more easily defined if the debate identifies appropriate approaches, in the Chinese context, to each of these three rather distinct activities. In particular, we see the National Natural Science Foundation (NNSF) as the appropriate body to finance curiosity-oriented research because the NNSF is able to apply high standards of peer review to individual projects. We will discuss later in this part of our report some options that we see for the support of strategic research (and the closely allied subject of precompetitive research), and in Part II we will offer a few comments on future support of big science.

Preservation and encouragement of basic research

Since the 1978 National Conference on Science and Technology, China's S&T policy has always contained explicit support for basic research, and this is equally true of the May 1995 Decision on Accelerating Scientific and Technological Progress. (However, many people would point out that the earlier declarations of support for basic research were frequently not carried through to actual resource allocations.) We have heard of, but been unable to quantify, an emerging problem in the basic-research system that will need to be tackled in the Ninth Five-Year Plan. Apparently, large numbers of the best of China's science graduates are being attracted by the higher incomes available in high-technology enterprises or in joint ventures, and so, it is argued, fewer and fewer are choosing careers in basic research. We know that there are programs at the national and provincial levels to attract young stars into research, including basic research. We suggest that it will be important to evaluate the success of these programs on an ongoing basis to ensure that an appropriate share of these younger talents continues to choose careers in basic research.

Although much attention in this report, as well as in China, is devoted to seeking improved means of linking China's investment in basic research to the long-term economic and social needs of the country, we agree with many of our Chinese colleagues who wish to retain three other important roles for basic research in a contemporary society:

- As a vehicle for advanced human-resource development (HRD);

- As a means of building a national knowledge base (a kind of social intelligence function to equip society so it can react to the uncertainties of a rapidly changing world); and

- As an expression of national culture in which knowledge is valued and the search for new knowledge is appreciated.

In all three of these roles, we agree with the emphasis placed by China on the importance of international collaboration — for all countries.

Continuing reform of the Chinese Academy of Sciences

The process of reform has been ongoing within CAS for a decade or more as CAS seeks to redefine the role of its extensive system of 123 research institutes. (The Academy's role as an honorific association of leading scientists is not challenged.) From a body that in the mid-1980s was almost 100% financed by an unconditional annual budget appropriation, CAS now has diversified revenue sources. Its annual income of about 1.4 billion CNY comes from the following:

- 20% from budget allocation from the national government;
- 30% from contracts with national ministries;
- 30% from contracts with enterprises; and
- 20% from contracts with provincial and municipal governments.

According to CAS, it has evolved away from an old Soviet model of an isolated set of basic-research laboratories with no real contact with either universities (which, before 1978, did little research) or enterprises into a system of national laboratories designed to

- Provide a national base of basic-research competence, across the natural sciences, in increasingly "open" laboratories that host visiting scientists from across China and around the world (this includes the Academy's special responsibility for China's activities in big science);

- Provide advanced training to talented young scientists;

- Participate in developing the most advanced high-technology sectors of Chinese industry; and

- Undertake research that is broadly defined as being in the public interest (such as environmental science).

CAS has developed a vision of its own future that would see its scientific work force decline substantially from its current level of 50 000[4] (out of a total of 90 000 employees). CAS would operate in a one-academy, two-systems mode: selected basic-research competence would be maintained using the government-supplied budget, and an increasing proportion of the work force would undertake applied tasks financed by external sources.

As with all such large and previously dominant institutions, CAS has many critics. We met groups who cited the May 1995 Decision as meaning that basic research should be done in the universities and that applied research should be done in enterprises — leaving little space for CAS institutes (as we set out in Part II, the Mission does not subscribe to this oversimplified interpretation of how to implement the May 1995 Decision). Additionally, we heard again the often-asked question, Why is the CAS engaged in agricultural research, given the existence of academies of

---

[4] We heard one suggestion that the number of scientists on staff should decrease to about 12 000.

Agricultural and Forestry Sciences and given CAS's apparent remoteness from any extension services?

The Mission believes that there is great scope for CAS to take the lead in defining new and productive relationships with the universities in its activities in fundamental science and with enterprises in the more applied aspects of its work.

The debates surrounding the CAS and its future should be seen, the Mission believes, as only part of a necessary and much wider debate on the restructuring and rationalization of the overall system of R&D institutes in China. Once there is in place a national social-security system capable of coping with redundant workers, the reform of many Chinese institutions will need to be undertaken as a matter of urgency.

Continuing reform of the research role of institutions of higher education

The range of research competence in China's more than 1 050 universities is truly great. At the top, some of the elite universities have wide competence across many fields of inquiry, and their scientists do well in national competition for funding. For example, from data provided to us by the State Education Commission (SEdC) and by the World Bank, of the 201 special institutions for research — national key laboratories, SEdC laboratories, and engineering research centres (ERCs) — now in place or whose establishment has been approved, a remarkable 100 (or 49.8%) are located at only 10 universities: Tsinghua (20), Beijing (14), Xi'an Jiaotong (9), Huazong University of Science and Technology (8), Shanghai Jiaotong (8), Tongji (8), Fudan (7), Nanjing (6), Jilin (6), and South East University (6).

More broadly, we were told that some 200 universities have the right to grant doctoral degrees and that 400 could grant master's degrees, but we suspect that in the majority of these institutions, the right to grant such degrees is probably limited to a few specific departments.

International experience suggests that doctoral-level training requires that the departments that offer doctoral degrees must themselves be heavily involved in research. The same is not necessarily the case for undergraduate- or master's-degree programs. This suggests that priority should be given in China to ensuring that all university departments authorized to award doctoral degrees become thriving research departments.

S&T information

The Mission did not visit any of the organizations in China that specialize in collecting and disseminating scientific information. Our views on this topic are therefore based solely on impressions gained from our discussions with policymakers and scientists. The impressions suggest the following:

- Accessing Western S&T literature remains a major problem for most research institutions. Only a few research institutes are able to purchase Western scientific journals.

- Whether to publish scientific papers in Western journals or in Chinese ones seems to be a hotly debated issue. Those in favour of publishing in Western journals argued that this enabled the Chinese research to be internationally peer reviewed and helped ensure that Chinese scientists maintained international standards. Those in favour of publishing in Chinese journals argued that their own journals were read and understood by many more Chinese scientists than had access to Western journals. Also, if the best papers were published overseas the quality of the Chinese journals would decline. Issues of national prestige and the ability of individual scientists to be listed in the Science Citation Index also featured in the debate. The Mission recognized the strength of the arguments on both sides of the debate. English is the international language of science, and as long as China seeks to be a part of the international scientific community then its scientists will be disadvantaged if they are unable to communicate in English. On the other hand, at this stage in its development, China needs to disseminate its scientific knowledge widely within China, and this means using the medium of the Chinese language.

- Beyond formal publications, much S&T knowledge is disseminated in Western countries through what is called gray literature, that is, literature and reports originally distributed among networks of colleagues without being formally published. This practice has been largely replaced in recent years by the dissemination of information through the Internet. In China we gained the impression that such informal reports are rarely distributed and shared. Information seems to be held secret by a department or agency, and there is a reluctance to share this with others. This impedes the dissemination of knowledge.

- The Internet is not yet widely used in China, but its benefits are well known and understood. Several of the scientists we met in eastern China have e-mail addresses. We anticipate Internet use will spread quickly, which could have a major impact on the diffusion of information and perhaps facilitate greater research collaboration. On the other hand, one of the implications of access to the World Wide Web is that it will further expose China to Western values and culture, not all of which are likely to be beneficial to China.

Underutilization of scientific societies

The Mission heard of the desire and capacity of many scientific societies in China to play a greater role in the S&T life of the new China. Only the Chinese Medical Society appears to have been accorded any significant operational role (in this case, screening drugs and assessing research directions and the quality of some applications for funding). It could be useful for SSTC and the Chinese Association for Science and Technology (CAST) to discuss opportunities for the greater involvement of scientific societies in the reform process.

Popularization of S&T

In the early 1960s the Chinese government invested substantial resources to popularize S&T (or the public understanding of S&T). Its reasons for doing this were in part to diffuse the scientific method of combining theory with practice in finding solutions to problems, to show that there were scientific explanations for phenomena previously explained by superstition, and to diffuse S&T knowledge. At the same time, the value of traditional knowledge was recognized, and considerable efforts were made to integrate traditional and modern knowledge.

Now, more than 30 years later, the state of public understanding of science is still an important policy issue, and for exactly the same reasons as in the earlier period. However, whereas most of the efforts of industrialized countries focus on popularizing science, the Chinese approach also recognizes the importance of spreading knowledge and understanding of how S&T effects people's lives.

The Mission believes that this is an important policy area for China and regrets that we did not learn more about the current programs.

Issues related to resource use

Environmental protection: an example of the problems of implementation

China faces staggering environmental problems related to its air, water, and land resources. Air pollution is a health concern in many industrial and urban areas: smoke and dust emissions are reported to be rising by 7% per year; sulfur dioxide emissions are expected to rise to 23 million t per year by 2000; and acid rain is becoming more widespread. Water resources are also in a state of crisis in many parts of the country: more than 40 of the largest cities are facing water shortages; and 100 million t of wastewater is produced per day, leaving some rivers dangerously polluted. Land resources are threatened by desertification, which reportedly reduces arable land by 2 100 km² annually; by industrial solid wastes, expected to increase to 250 million t per year by 2000; and by the widespread use of chemical fertilizers. Unless these trends are reversed, China is expected to become the largest single producer of both carbon dioxide (the leading cause of global warming) and sulfur dioxide by the next century.

The problem of protecting the environment while maintaining economic and social development is clearly a major policy issue at all decision-making levels in China, and the Mission was impressed with the concerns expressed and the initiatives under way. The National Environmental Protection Agency is collecting statistics on the equally staggering costs to the economy of environmental degradation, which add up to some 100 billion CNY annually.

The main approaches being taken to protect the environment are the following:

— Enacting tough new environmental legislation (including even the death penalty);

— Imposing fines for pollution, based on the polluter-pays principle;

— Requiring environmental assessments and imposing more stringent standards for new industries;

— Establishing industrial development zones, where environmental-protection infrastructure, like waste and sewage disposal, can be shared and emissions can be more stringently controlled and monitored;

— Investing in cleaner technology, land rehabilitation, and afforestation projects;

— Creating major infrastructure schemes for energy development, especially hydroelectricity and water transfers; and

— Improving public environmental awareness and education.

The government has strengthened its environmental-monitoring and enforcement capability. By the end of 1992, a network of 1 808 monitoring stations had been established under the authority of national, provincial, county, or township administrations. These monitoring stations have a total staff of more than 25 000, and more than 16 000 staff are employed in environmental regulatory agencies. Over the last decade, China has put in place a comprehensive environmental-management system. Why then did the Mission hear concerns from national, provincial, and township officials about the implementation of environmental-protection measures?

The issues are reported to relate to

- The regulatory system for pollution control;
- Low prices for resource inputs (particularly for energy and water); and
- Negative environmental impacts resulting from other S&T policies and programs and broader economic and social policies.

Current debates about environmental regulatory policy centre around the following:

- Pollution charges are too low: it is cheaper for polluters to continue to pay the fines than to clean up;
- Retrofitting polluting equipment is expensive, and there is probably a need for economic incentives, such as subsidies for installing clean technology or tax rebates;
- Polluters pay only for the worst-offending pollutant, so there are few incentives to reduce emissions of other pollutants;
- Fines collected may be allocated to environmental protection or general revenues but are not usually reinvested in improved and cost-shared treatment facilities;
- The monitoring capacity of the environmental-protection agencies (EPAs) requires considerable strengthening in technology, trained staff, and coverage of the monitoring network; and
- TVEs are less effectively regulated than SOEs but account for a significant, and increasing, proportion of environmental pollution.

International experience

China has already adapted international experience in standard setting and monitoring to its own situation. In some respects, China is an innovator: for example, revenues from pollution fines are allocated to local funds for distribution to enterprises as subsidies or soft loans to finance cleaner production technology. Future policy reforms under consideration in China have been tried in a number of Organisation for Economic Co-operation and Development (OECD) countries: these include pollution charges that are high enough to encourage firms to invest in cleaner production technology; and a more integrated system of economic instruments, including not only pollution charges but also taxes, tradable emissions, and subsidies.

Resource pricing

One of the emerging policy debates in China is on resource pricing. The Mission was told that, in general, the prices of energy and water inputs are much lower than the real cost of supply and that this leads to resource inefficiencies, overproduction of wastes, and increased pollution. The Mission would suggest an even broader policy debate: on developing a natural-resource accounting framework and on measuring the value to the national economy of environmental goods and services. China's environment provides not only all its natural resources but also its waste-management services — waste products are diluted, absorbed, and removed through air circulation, stream flows, and biogeochemical cycling. In addition, biodiversity has an immense value to the Chinese economy, as a source of genetic material for improved food species and as a source of new products and pharmaceuticals, as well as added value to the tourism industry.

It is generally recognized that official coal and water prices are far too low in China. This has negative repercussions in the case of coal mining because low coal prices discourage coal mines from investing in environmental-protection or safety measures. The recent expansion of township enterprises into coal mining has exacerbated the environmental damage caused by the mines themselves. More generally, the low price of energy provides a perverse incentive for industry to pollute.

The total (economic, social, and environmental) costs of coal include not only the mining costs but also transportation costs from northern and western China to the populous east and south. Currently, 40% of China's railway capacity is reserved for coal transportation, but transportation is a main bottleneck to supply and also incurs additional pollution costs. Washing the coal with water would reduce the amount of coal that is transported, as well as the ash produced during burning, but in 1985 only 16% of all raw coal was washed, compared with 80% in OECD countries. In addition, converting coal into electricity incurs costs.

The policy debate is complex, going far beyond the acceptance that the price of energy be increased into a broader examination of alternative policy instruments, from pollution fines to consumption taxes and tradable permits. The cost of coal, for example, could be increased at the mine head by requiring mining enterprises to pay a levy for the costs of environmental damage caused by extraction. Reforming the transportation component in the cost of coal would inevitably involve a wider consideration of freight rates and regional inequalities in distance to energy users. Incorporating the total costs of energy consumption in the price to consumers would involve a policy equation that includes comparisons of the locational advantage of sites for power production and local environmental-quality standards.

Similarly, the low government-regulated prices that farmers receive for much of their grain production means that they cannot pay higher prices for water, which is provided to them at costs below the cost of supply. There is little incentive for farmers to invest in water-saving equipment or processes. The negative environmental outcome is increased abstraction of groundwater and surface water, which lowers water tables and reduces the minimum surface flows necessary to provide environmental services, such as pollution flushing, and the maintenance of floodplain vegetation and aquatic resources. The result is that the farmers waste water, but the policy reform needs to encompass not only water pricing but also grain prices.

An analysis conducted for the China Council of the 1993 water prices in Beijing showed that for recovering the total costs for supplying water in Beijing, prices should be 10 times higher for industrial users, 15 times higher for municipal users, and 77 times higher for agricultural users. If prices were raised sufficiently to reduce current waste (more than 50% of water in Beijing is still used for agriculture) such that the huge costs of the Eastern Route Water Transfer Project could be delayed even by a few years, the savings would be enormous.

The Mission believes that the resource-pricing policy debate is a very important one for the S&T community in China to address. It is also evident that any policy reforms will need to go far beyond S&T and encompass broader economic and social reforms.

Rural development

During the decade 1985–95, S&T reforms contributed to the great changes taking place in the Chinese countryside, but they did so in a broader context of social and economic change. A number of national goals have guided S&T reforms for rural development:

— Ensuring national food security, especially self-sufficiency in grains;

— Providing low food prices and greater food availability for the poor;

— Increasing agricultural productivity as land per capita decreases;

— Increasing rural incomes and rural employment;

— Transforming agriculture from a subsistence to a commercial basis; and

— Maintaining rural society by reducing the rural–urban prosperity gap.

Eighty percent of women live in rural areas, and women constitute 70% of the illiterate adult population. In addition to the goals listed above, increasing the access of girls and women to education and to the benefits of S&T, as well as strengthening their participation in economic activities, is clearly critical to rural development.

Two groups of S&T reforms have had a great impact in rural areas; these are the priority given to agricultural research and the contribution of the Spark Program (discussed below) to the rapid development of rural industries. Both groups of reforms have brought both successes and new challenges.

Agricultural productivity

A key goal has been to apply S&T to raise agricultural productivity. This led to some early successes: for example, the net result of technological improvements and the changeover to the family farm was an increase in grain productivity of 43% between 1978 and 1984; for wheat, the increase in yields for the same period was 61%. These spectacular trends have not been maintained, although there have been further technological developments such as improved genetic varieties and the use of chemical-intensive production methods. It has been estimated that agricultural chemicals have accounted for about 30% of the increased grain production since the 1970s and have reduced the labour required per hectare. Despite this, there is a lower efficiency in the use of fertilizers than in other countries. Furthermore, there is widespread environmental contamination because of heavy chemical loads in agricultural runoff. The costs of inputs, such as chemicals, have increased more rapidly than the prices paid to farmers for their outputs, especially grains.

The Mission was told that there is concern about the dependence of Chinese agriculture on chemical inputs, both because of their cost and because of their negative impact on the environment. There are renewed calls for an agricultural transformation that is based on biological pesticides and organic fertilizers, as well as on plant varieties that are drought resistant and adapted to marginal land. These are priorities for S&T investments. However, we were also told that the present combination of economic and technological developments has created a situation in which farmers are discouraged from growing grain and are turning to more profitable activities, such as growing alternative crops, including vegetables; working in township enterprises; or migrating to urban areas. These are issues that clearly go beyond the domain of S&T.

Rural enterprises

A major contribution of S&T reform to rural development has been the Spark Program. This successful combination of packages of proven technologies and an effective delivery system, including extension and demonstration projects, has been hailed as a major success. The Mission was impressed with the projects we visited, both those focused on agricultural postharvest storage and processing and those less directly based on agriculture. The Spark Program has shown that S&T can be packaged to benefit the lives of the rural poor. Other countries could benefit from a close look at the Chinese experience. It was not clear to us how far the Spark Program included specific consideration of the needs of rural women or, indeed, whether it has led to greater benefits for men than for women. It clearly has potential to improve the status of rural women by increasing their incomes and their participation in rural enterprises.

The proliferation of rural enterprises has become a greater generator of rural prosperity than agriculture itself, and this has led to some of the problems noted above. Between 1980 and 1989, agriculture's contribution to rural productivity rose from 192 billion CNY to 653 billion CNY (an increase of 240%); rural industry's contribution rose from 54 billion CNY to 589 billion CNY (an increase of 991%). The rural enterprises have created employment for the increasing numbers of surplus rural labour, but they raise concerns about health and safety at work, as well as contamination of the rural environment. They also serve to reduce the attractiveness of agriculture, which is still the key to China's future economic prosperity.

Future reform of agricultural R&D

The national and provincial agricultural R&D institutes have appeared to be slower in taking advantage of the opportunities presented by reform, and the development of an effective technology market for much of their research output has also been slow. The reasons behind these observations, we believe, lie in the inexperience of the research staff in dealing with technology-market conditions and the reluctance of end-users to pay for technology and consulting services that had been provided at no charge in the past. The scope for more effective technology transfer is improving with the growth of rural enterprises, which are providing a new market for S&T, especially in the areas of postharvest and food-processing technologies. This new demand should help to initiate the necessary changes in the research programs and priorities of the agricultural R&D institutes.

The urgent challenge facing policymakers concerned with the improvement of S&T activities in the agricultural sector will be to restructure the existing research service to make it more relevant and efficient and more capable of responding to the urgent needs of the rural sector in the next decade and beyond.

Issues related to the financing of S&T activities

Target for R&D spending for 2000

The May 1995 Decision contains an ambitious target for the growth of China's gross expenditure on R&D (GERD). It is proposed that GERD be increased in a 5-year period from its present value of 0.6–0.7% of gross domestic product (GDP) to the target of 1.5% of GDP (Figure 2), with enterprises accounting for 60% of the target and with the governments of "economically developed regions" expanding their expenditures at a higher than average rate.

The Mission believes that rapid consideration needs to be given to the adequacy of the existing array of fiscal and tax incentives currently offered to promote R&D activities within enterprises if the government is going to be able to convince enterprises to change their behaviour. Some consideration might be given to general tax incentives for R&D activities, such as those offered in some industrialized countries (for example, Australia and Canada). An additional route to increasing R&D funding worth exploring could be the imposition of levies on production, a system used to good effect in supporting agricultural research in many countries.

An additional factor that needs to be taken into account is the influence of the rate of inflation on the decision-making behaviour of enterprise management. In industrialized countries, periods of high inflation appear to induce a stance of "risk aversion" in many enterprises, which

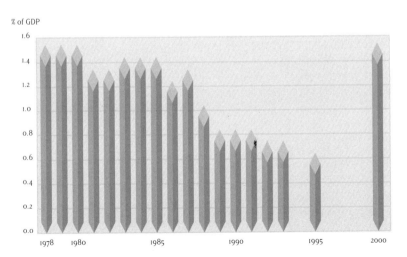

Figure 2. China's gross expenditure on R&D as a percentage of gross domestic product (GDP).

take a conservative view of investments in R&D during such periods of economic uncertainty. Conversely, periods of rapid expansion in industrial R&D expenditures tend to appear in periods of low inflation. As a consequence, the attainment of the goal for S&T spending may be highly dependent on China's macroeconomic performance between now and the end of the decade.

Chapter 3

EMERGING ISSUES

Policy advice

In the market economies of industrialized countries, scientific research and technological development are widely distributed between the public and private sectors, among government institutions, universities, and enterprises. This has led governments to seek advice on new policy directions from people engaged at these different kinds of institutions, as input to decision-making, which remains the prerogative of government. Each government has institutionalized the process of advice-receiving in a form appropriate to its system, but all systems share in common the idea that senior political leaders should be able to receive such direct inputs to their work. As China's socialist market economy evolves, the Government of China should give thought to how it can tap into the growing experience of enterprises, universities, and R&D institutes as it continues to evolve policies for the promotion of innovation and technological change in the light of the ever-changing global economic system.

Need for policy integration

The countries of the industrialized world have now all accepted that technological change is a principal driving force within their economies. As a result, they have concluded that they need to pay special attention both to policies to promote S&T and innovation and to the integration of the main elements of these policies into the other principal elements of their public policies in other fields. This has meant paying attention to the level of coherence that can be achieved between economic, trade, education, defence, and other policies and those designed to promote technical change and innovation. The Mission did not get a clear picture of how this policy integration might be promoted among the main commissions that set policies at the highest level in each level of government (national, provincial, and municipal), and we suggest that this needs some clarification.

We are aware of many interpretations on how the proposal in the May 1995 Decision on Accelerating Scientific and Technological Progress (Box 2) to create "a leading state science and technology group ... to strengthen overall policy making and management related to scientific and technological work throughout the country" will be implemented. Any such leading group could certainly help by looking into the question of policy integration.

Setting research priorities

The impression gained by the Mission was that priorities for research, both short term and long term, were set following debate among groups of scientists. This process is highly appropriate for priority setting within and between scientific disciplines when the choices depend on purely scientific criteria. When priorities need to be set for strategic research — that is, research that might have economic or social benefits in the long term — then a mix of scientific, economic, and social criteria need to be used.

In recent years, the more-industrialized countries have developed quite sophisticated priority-setting techniques in which criteria relevant to market considerations are integrated with scientific criteria. Governments have an important role to play in facilitating these priority-setting exercises, although members of the scientific and business communities also play important roles in their conduct.

The term *research foresight* is often applied to these activities, and different countries have evolved techniques appropriate to their own domestic situations. The Japanese were early pioneers in the development and application of these approaches; more recently, the UK government has facilitated a major foresight study in the United Kingdom; the South African government is about to launch a foresight study in South Africa; and Australia has also developed some original approaches for setting research priorities.

Box 2

The Decision on Accelerating Scientific and Technological Progress (6 May 1995)

The Decision on Accelerating Scientific and Technological Progress (see *National Affairs* 1995) is based on 11 major points, some of which are supplemented by a series of principles. The major points are the following:

I   Implementing the idea that Science and Technology are primary productive forces in all fields;

II   Energetically push forward scientific and technological progress in agriculture and rural areas;

III   Improve the quality and efficiency of industrial growth through advances in science and technology;

IV   Develop high-technology and its industries;

V   Promote scientific and technological progress in social development;

VI   Firmly tighten basic research;

VII   Continue to restructure science and technology management and establish a new system of science and technology management, compatible with the socialist market economic system and the law of scientific and technological development.

VIII Train a contingent of highly qualified scientific and technical workers and enhance the whole nation's scientific and technological level;

IX   Increase science and technology inputs through various channels and at different levels;

X   Further opening up China to the outside world and extensively launching international scientific and technological cooperation and exchanges;

XI   Effectively strengthening Party and Government leadership over scientific and technological work.

Given China's commitment to investing in research that will yield benefits to the Chinese economy and society within 10–15 years, the Mission believes that some of these more sophisticated priority-setting techniques might be usefully applied in China.

A national system of innovation

In industrialized countries, the focus of policy development that is concerned with the role of technological change in the economy is now firmly on innovation policy, and the concept of an NSI has evolved.[5] Although we have seen occasional references to innovation in Chinese policy pronouncements, it is our impression that China is still highly focused on R&D policy — an important subset of innovation policy, but not as comprehensive a concept. We believe that China should now turn some of its attention to the NSI mode of analysis as a means of identifying the future needs for reform in the S&T system and in the S&T system's relationship to overall economic and social activity in the country.

We have included in Part II a first attempt to identify the outlines of China's NSI. In our approach, we place emphasis first on identifying the functions of an NSI and then on identifying the stakeholders whose interests are affected by the system. Any analysis of the functions of an NSI (see Table 2) needs to take into account the following:

- Policy and resource-allocation functions;
- Regulatory functions;
- Financing functions;
- Performance functions;
- HRD and capacity-building functions; and
- Infrastructure functions.

The groups of stakeholders include

- Policy-making institutions;
- The principal S&T institutions;
- New organizational forms created by the reform process;
- Organizations of the scientific community;
- Relevant financial institutions; and
- Regulatory bodies.

---

[5] A definition of *innovation* is given in the first paragraph of Chapter 5.

Models of R&D and innovation

Many people with whom the Mission met appeared to base their ideas on a simplified linear model of R&D and innovation. This model, which is usually referred to as the "technology-push" model and has been largely abandoned in the industrialized world, suggests that these processes are simply linked in some variation of a scheme like the following:

Basic research ➤ applied research ➤ technological development ➤ innovation

The practical consequence of belief in this model is that institutions and programs tend to be situated at discrete points along this spectrum, specializing in one or other activity, and tend to be organized along lines of scientific discipline (for example, institutes of basic research in some aspect of physics).

Other models of the innovative process that have been developed seek to incorporate the ideas of market pull and of the need for interaction among a whole range of technical activities that are seen to contribute to innovation in a modern enterprise. The latest attempt to understand the innovative process — and hence to be able to prescribe ways of enhancing the innovative performance of enterprises or institutions — is the system integration and networking model. An Australian review of innovation studies (Tegart 1995, p. 1) underlined the point that current models of the innovative process

> emphasise that innovation should be viewed as a team effort, with formal and informal networking as the crucial factor in transferring knowledge amongst participants. These models recognise that there can be both long-term and short-term outcomes of the innovation process and that different technologies induce different patterns of innovation and diffusion, including new interdisciplinary groupings.
> The [so-called] fifth-generation innovation process [Figure 3] identifies the need for firms to be systematically innovative in all of their activities. This includes their linkages with customers, suppliers, information sources, research providers and all the various parts of their networks. In particular, the growing interdependence of national economies — so-called globalization — means that these linkages must be global as well as national.

The reference in Figure 3 to the value chain is to the technological trajectory followed by innovative enterprises as they continually seek to introduce greater added value to the products or services they offer on the market.

In industrialized countries, much attention is paid to fostering the creative interactions among people doing research, design, and production, and this has led to the creation of new groupings of scientists and engineers — in networks and consortia — and to significant investments

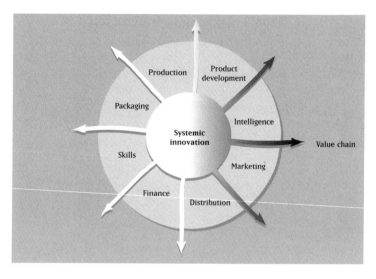

Figure 3. The so-called fifth-generation innovation process.
Source: Tegard (1995).

in research at the frontiers of science in areas considered to be likely sources of future technological development of economic or social importance (strategic research). China should carefully examine the utility of this experience as it moves to further institutional reform in its S&T system.

Role of national key laboratories, ERCs, and national research centres

The S&T-reform process has given rise to three institutional innovations that have been superimposed on the existing extensive and complex organizational structure of Chinese S&T. These innovations were the following:

– The designation of a series of national key laboratories — This program began in 1984, and today 155 such laboratories have been identified, 50 of which are located in CAS institutes. Selection of laboratories to be so designated involved a peer-review process organized by NNSF, and the final decisions were jointly made by SSTC and SPC;

– The establishment of an extensive series of ERCs, proposed as vehicles for improving the transfer of domestic and foreign technology to enterprises[6] — To date, SSTC has established 25 out of a planned 56 ERCs, and the World Bank has given approval for the financing of 46; and

— The establishment, during the Ninth Five-Year Plan (1996-2000), of a series of at least 10 national research centres, which appear to come in three types:

- The geographically distributed network, in which institutes and laboratories in different locations form linkages among their activities,

- The consortium in a common location, in which different organizations (national key laboratories, university institutes, CAS institutes) join together in a common effort to define a joint research program based on the strengths of the different participants, and

- The new centre in new facilities, a costlier version that involves investment in new physical plant.

The original motivation for the designation of national key laboratories appears to have been a desire to consolidate, and to ensure continued governmental funding for, a series of the most productive laboratories engaged primarily, but not exclusively, in basic research, thus ensuring that national competence is maintained at the frontiers of research in what are perceived to be important disciplines. The Mission did not have the opportunity to explore the systems of governance of these key laboratories, so we are not able to comment on the provisions that have (or have not) been made to involve external peers in processes to help the laboratory director maintain standards of scientific quality.

The various ERCs appear to have been selected in a competitive process, with individual institutions of many types making proposals to the sponsoring commissions. This approach has the merit of basing the ERCs on established technological capability but gives rise to questions about the extent of national coverage that will be achieved, even with more than 100 such centres eventually being set up. The Mission could find no evidence of any concern to understand the geographic dispersion of the centres — are they located in places that give them easy access to the enterprises most likely to need their services? What kind of outreach capacity will the centres have to take their services to potential clients? What

---

[6] According to the World Bank (1995, p. 15), "the ERCs would be market-oriented independent limited companies under Chinese Law, with the objective of adapting, developing and diffusing technologies, in particular those that have positive environmental impacts. The ERCs will produce, under contract, equipment and systems designs, prototypes and customised products and services that adapt foreign and domestic technology to local inputs, scales of production, and other market conditions."

kind of networking will be promoted to allow the set of centres to respond quickly to client needs?

For the medium term, a larger question arises. Will these ERCs, at some future date, be elements of a national system of technical support to Chinese enterprises? If such a system were to emerge, how would it be organized? How would its activities be financed? What range of services would it offer? From experience in both industrialized and developing countries, the five most common needs of small- and medium-scale enterprises (of which there must be many in China) are the following:

- Access to finance [and China has already a variety of schemes at the national, provincial, and municipal levels];
- Access to management training;
- Access to market information and to markets;
- Access to skills upgrading for employees affected by technological change in the workplace; and
- Access to best-practice technology, with best practice defined in terms of the acquiring enterprise's capacity to absorb the technology in question.

There is rich international experience in providing technological assistance to companies, both in the industrialized and in the rapidly developing countries, and SSTC might consider investigating the applicability to China of some of the experiences of programs such as

- The Manufacturing Extension Partnership, operated by the National Institute of Standards and Technology in the United States;
- The Industrial Research Assistance Program of the National Research Council of Canada;
- The Technology Adoption Program of the Singapore Institute of Standards and Industrial Research; and
- The technology-commercialization activities of the Fundación Chile.

In each of the four examples cited, the program management places great emphasis on hiring staff with many years of industrial experience. We surmise that the Chinese ERCs will initially be staffed by former researchers from within the sponsoring institutions, so we would caution ERC management to watch carefully to identify those staff members who have a genuine understanding of the operating environment within enterprises. The enterprises that these staff members will advise all have to face the daily challenge of the market; the advisers need to understand the nature of that challenge.

The introduction of the concept of the national research centres, as the Mission understands them, offers opportunities to involve the most progressive of Chinese enterprises in the new forms of joint activity now common in industrialized countries. Much strategic research (that is, research at the frontiers of knowledge in fields of S&T believed to be important to future social or economic performance) is now carried out by networks or consortia of R&D institutions, and in most countries these consortia include government research institutions, universities, and enterprises. If China takes steps to involve enterprises in the new national research centres — both in their governance structures and in their research teams — it would be creating powerful institutional means to draw on the traditional strengths of the national key laboratories and the emerging skills of the ERCs in creative ways to focus enterprises' interests in their own long-term development. This, the Mission believes, is an opportunity worth exploring.

Good examples of new forms of collaboration are to be found in

- The Australian Co-operative Research Centres Program, funded by the Australian Commonwealth Department of Industry, Science and Technology;

- The Canadian Networks of Centres of Excellence Program, funded by the Natural Sciences and Engineering Research Council of Canada; and

- The industrial R&D consortia operated under the US federal *National Cooperative Research Act* of 1984.

Human-resource issues

It would be impossible for an international mission to ignore the enormous problem bequeathed to Chinese science by the Cultural Revolution. For almost 10 years, between 1966 and 1976, not only was scientific research, especially agricultural research, in disarray, but also scientific education in schools and universities was severely disrupted. There must be few scientists over the age of 40 whose training and research were unaffected.

We were impressed with the extent to which this legacy is recognized in China today and by the steps that are being taken to remedy it. The measures mainly call for retirement of older scientists; transfer of others from research to routine activities; some retraining of middle-aged scientists; and special inducements, rewards, and early promotions for younger and middle-aged scientists with expertise.

There also seems to be a long-term view on the brain drain. When the open-door policy was introduced, many Chinese scientists were sent overseas for further education and training. Many decided to remain overseas, especially following the events of June 1989. Despite this loss, China decided to maintain the flow of scientists going overseas. Today that policy seems to be paying off. The Mission heard of many younger scientists who have returned to senior positions. Others are encouraged to return to give lectures or to develop collaborative research programs involving Chinese institutions and their own foreign institutions.

No other country has been faced with such a legacy, which at the same time provides huge opportunities and responsibilities for the younger generation. How well the aftermath of that legacy is managed will be one determining factor in how well China can implement its ambitious policies.

Some benefits and costs of the S&T reforms for research institutions

From the interviews with individuals working in research institutions and universities, the Mission drew up the following very tentative balance sheet of benefits and costs of the S&T reforms of the last decade.

Some benefits of the reforms

- They provided mechanisms to permit the best researchers to concentrate on research and to allow others to be usefully employed in "spin-off enterprises."
- They increased mobility of researchers, which allows top institutes to attract top scientists.
- They required institutions to find new ways of linking their work to societal needs.
- They provided peer-review mechanisms for concentrating resources in the best groups.
- They put heavy emphasis on encouraging talented young and middle-aged scientists.
- They provided more opportunities for returning Chinese scientists.

Some costs of the reforms

- They led to commercially unskilled scientists trying to become entrepreneurs and enterprise managers, with, as a consequence, many enterprise failures.

- They led to a move of bright young scientists away from basic research because financial rewards are potentially higher for applied research, which has created more opportunities for employment with new enterprises.

- They led to rewards within institutions being preferentially for those who can "sell their services and technology." This increases competition between departments and leads to short-term research and major difficulties in promoting interdepartmental and interdisciplinary research.

- They led to the directors of many institutes having little opportunity to direct because individual researchers exploit the market to find funds to support their own research.

Some key unresolved issues

- There are still too many government research institutes.
- Too many institutes are overstaffed.
- Some important new tendencies in the global organization of S&T have still not been fully grasped or institutionalized in China.

Some final observations

We began this report with comments covering the main impressions that we gained during our visit to China in November. As we finish it, we wish to add a few further observations, more oriented to our impressions of useful future directions for China to consider pursuing in the light of experience in other parts of the world.

We share the view that S&T is the driving force behind contemporary economic development and that S&T has great potential to contribute equally to social development, provided that its development and application are carefully managed.

Given China's enormous population and the limitations that it faces in its available arable land, we agree that high priority has to be given to agricultural development through R&D; the two major concerns in this area to which China should pay attention are the need to have policy statements backed up by appropriate budgetary expenditures, both by governments and by enterprises, and the need to turn the large array of existing agriculture-related S&T institutions into an efficient and effective system.

To tackle the problems of accelerating industrial growth through technological innovation, China's NSI needs to evolve into an interactive system, ending the tradition of institutional isolation. There is also a need to pay greater attention to the effective assimilation and mastery of imported technology, an essential precursor to having enterprises become a continuing source of innovative activity.

Becoming an increasingly innovative society requires that government create a policy environment that fosters creativity and investment. This in turn requires that all of the principal policy-making bodies of government work together to consciously integrate the many elements of government policy into the desired whole.

The Mission is in favour of the policy of continuing to support research at the forefront of scientific knowledge. We believe that China has the capacity not only to undertake ventures in big science but also to develop a strong tradition of strategic research that is closely linked to China's short-term programs of technological development. For this, some institutional realignment will be needed. A look at international experience in designing research consortia should also be helpful.

We are strongly in favour of the openness that has been developing in China's S&T system over the last decade, and we see the exposure of young Chinese scientists to foreign ways of doing and managing science as being crucial to China's own efforts.

Part II

Detailed Observations of the Mission

Chapter 4

GENERAL OBSERVATIONS ON THE REFORM PROCESS

The task facing the Mission was daunting: in 3 weeks, it was to assess the results of S&T reforms that had been implemented over more than a decade, in a country with more than four million scientists and engineers and 20 000 S&T institutions. Furthermore, many of the fundamental institutions of China's economic and social system were undergoing change.

It soon became evident to the Mission that the utility of its views of what was happening to S&T in China would be profoundly conditioned by its understanding (or lack of understanding) of what was happening in the overall process of change in the country. We therefore feel it necessary to summarize our knowledge of what is happening in the reform process in general, how the roles of some key institutions — particularly the operational ministries of government — are changing, and what added complexity springs from the current state of national–provincial–municipal relations in the fields of S&T. We devote this chapter to our views on these important background issues.

Evolution of the reform process

China's S&T system emerged from the Mao Zedong period with many problems. By the end of the Cultural Revolution, research institutes, universities, enterprises, and all other "work units" in urban society had become cellular enclaves of work and social services that were resistant to any forms of coordination and cooperation. Research agendas tended to be stagnant, there were no possibilities for professional personnel to move to better opportunities or to positions better suited to their knowledge and skills (they were regarded as "unit property"), and productivity gains in the economy were falling.

Meanwhile, outside China, the world's S&T had developed rapidly during the years when the Cultural Revolution's radical politics occupied China's national attention. The market economies of the capitalist world, including those of China's Asian neighbours, were increasingly basing their further growth and development on productivity gains brought about by technological and managerial innovations ("intensive growth") while China and others with centrally planned economies continued to rely on the mobilization of new sources of labour and capital inputs to expand economic activity ("extensive growth"). It was appropriate, therefore, that as the post-Mao period began, it did so with the major National Conference on Science and Technology of 1978.

This conference marked the beginning of the S&T-reform process, which still continues today. Although the conference attempted to set a broad and ambitious research agenda for basic science and high technology (an agenda that was postponed and then abandoned within a year of its announcement), the conference's greatest significance was ideological. In an attempt to reverse the low status of science — and of scientists and engineers — that had resulted from onslaughts against technical intellectuals during the Cultural Revolution, Deng Xiaoping's speech at the conference celebrated the roles of scientists and engineers in society by making it clear that S&T was the principal "productive force" (and not part of the "superstructure"). Scientists and engineers should therefore be regarded as part of the working class and not somehow politically suspect as they had been since the Anti-rightist Movement of the late 1950s. The Four Modernizations policy was reaffirmed, and the modernization of S&T was seen as the basis for the modernization of agriculture, industry, and national defence. From this basic ideological change came a gradual improvement in the political standing of technical intellectuals, including the gradual replacement of political cadres in leadership roles in research institutes and institutions of higher education by qualified scientists and engineers.

By the early 1980s, the ambitious research themes of the 1978 conference had been replaced by a more practical orientation of having S&T serve economic development and having economic development be based on S&T. This orientation, of course, also continues today. However, the pursuit of this objective in the early 1980s required that the chronic separation of R&D from economic production be overcome, that the operational mechanisms and management principles of both R&D and the economy be changed, and that scientists and engineers have the professional mobility needed to ensure that their talent was used appropriately. For both research institutions and industrial enterprises, which had long been comfortable in receiving their annual appropriations without concerning themselves with national innovation needs, the switch to the new orientation would require major reforms — reforms in the economy as a whole, in enterprise management, and in the S&T system.

Various reforms in the S&T system began in the early 1980s on an experimental basis, against a background of aggressive acquisitions of foreign technology to upgrade quickly the technological levels of Chinese industry, the early rural reforms, and the beginnings of reforms in the urban economy. Perhaps the most important of these S&T reforms was a series of measures intended to change the ways research was funded. The old system of guaranteed annual appropriations from the state served neither basic research nor applied R&D well: it provided for no effective mechanisms of accountability, either to peers for the quality of basic research or to the economy for applied work that required economic justification.

In the face of increasing exposure to foreign governmental and corporate models for organizing and managing S&T, China began to encourage new multichannel approaches to funding. These included a trial research "foundation," run by the CAS, in which investigator-initiated proposals for support — from both CAS and non-CAS researchers — would be subject to peer review before funding. This CAS program was then superseded by NNSF when it was established in 1986. The Mission was impressed throughout its visit by the high regard with which NNSF and its programs are held and by the fact that many in the Chinese technical community commented on the positive effects this new approach to research management has had. Indeed, throughout the 1980s, as new national research programs (such as the 863 Program) were launched, the principle of peer review was embraced by policymakers, administrators, and researchers as a critical element in sound research management.

In addition to the new "foundation" approach to funding, other approaches were introduced. In particular, efforts were made to encourage the flow of funds through "horizontal" channels, especially between research institutes and enterprises and between research institutes and local governments. Institutes were encouraged to do contract research and to sell their research results. It soon became clear, however, that for these measures to work, other important changes in policies and attitudes would be required. Research institutes and institutions of higher education had to be induced to take the trouble to enter into contracts and to market their results. To this end, the guaranteed annual state appropriations to institutes deemed to be doing work of an applied nature were gradually reduced. But once the elements of marketization were introduced, it also became clear that if technical knowledge was to be considered more of a commodity, it would be necessary to find a means to place a value on that knowledge and to clarify and protect the ownership rights of this commodity. The establishment of China's patent system in 1985 represented a step toward meeting the latter objective.

The March 1985 Decision on the Reform of the Science and Technology Management System represented an effort to sum up and formalize many of the reform experiments of the early 1980s and to push the S&T-system reform to higher levels. Key concepts in reform thinking throughout the 1980s that were incorporated in the Decision were that market mechanisms should be substituted for administrative ones in the S&T system, where possible and appropriate, and that S&T reforms should be harmonized with the broader ongoing economic and enterprise-management reforms.

The Decision and subsequent laws and policies unleashed a whole range of reform initiatives, which have

- Changed the ways S&T is thought of in society (involving, in particular, the idea that technology can be treated as a commodity that can be bought and sold in the market);

- Changed the ways S&T activities are funded; and

- Stimulated a range of institutional innovations to exploit market opportunities for technologies.

The latter have included innovations that expanded the range of ownership arrangements of institutes and technology-oriented enterprises from that of the government-owned type to collective and private forms.

Of particular interest were the efforts made to further stimulate the relationships between research and production by

- Clarifying the state's determination to reduce or eliminate traditional "vertical" sources of funding;

- Encouraging and formalizing technology markets; and

- Encouraging the merging of research institutes and enterprises.

By the mid-1980s, partly as a response to these initiatives, China was seeing efforts by research institutes, institutions of higher education, and individuals to commercialize technical knowledge through the establishment of new companies. These were established for the merchandizing of technology-intensive equipment, for technical services, and in some cases, for the introduction of new products. In addition to serving the reform objective of commercializing research, these NTEs provided a means for pursuing other reform objectives, including finding alternative sources of funds for research units, increasing the mobility of scientists and engineers, and helping research units reduce the numbers of redundant staff they had accumulated.

The successes of these reforms during the late 1980s and early 1990s have been mixed. With new sources of funds from innovative state projects and programs and with horizontal flows of funds beginning, capable and aggressive institutions and individuals were able to actually increase their overall revenues, despite the loss of some or all of their core state funding. Others, however, saw their absolute amounts shrink and, because of their fields of study, lack of entrepreneurship, or lack of quality, began to experience very hard times.

The efforts to establish strong horizontal links between research units and enterprises also had mixed results. Research contracting and technology markets have developed, but both institutes and enterprises have at various times and in various ways been dissatisfied with these mechanisms. For instance, enterprises felt that institutes were not providing packages of technology that could be directly used in production, and institutes felt that SOEs still had not developed an appreciation for the value of technological change. Differing approaches to placing a value on technical knowledge and to distinguishing between codified and tacit knowledge have also created difficulties. Interestingly, with the growth and economic success of collectively owned TVEs, the research institutes gradually began to find a greater demand for their knowledge and skills among these new types of enterprises than they did with the SOEs. In addition to entering into contractual arrangements with institutes for the transfer of codified knowledge, TVEs have hired scientists and engineers away from research centres, a practice that has led to IPR disputes in the courts. Efforts to merge institutes and enterprises have also run into difficulties because of problems of technological-asset valuation, differing organizational cultures, and inability to accomplish such mergers without one side having to shoulder the social-services burdens that the new partner would inescapably carry with it.

Meanwhile, as reforms in the operational mechanisms have proceeded — with both accomplishments and problems — new national programs have been introduced that both focus on national economic and research objectives and seek to employ and promote the new operational mechanisms. Among others, these include the National Program for Tackling Key Technology Problems (1984), the Spark Program (1986), the National Development Program for High Technology, or 863 Program (1986), the Torch Program (1988), and the Climbing Program (1991). The distribution of these programs and their emphasis on commercialization of technology are shown in Table 1 (the simple classification used in this

Table 1. Orientation of new national research programs in China.

Basic research	Applied research	Extension or commercialization
National Key Laboratory Program	National Program for Tackling Key Technology Problems	National Program for Key Industrial Experimental Projects
National Funds for Natural Sciences		Industrial Technology Extension Program
		Spark Program
		National S&T Achievements Spreading Program
Climbing Program	National Development Program for High Technology (863 Program)	Torch Program
		Trial Production and Appraisal Program

table follows the linear model of R&D that the Mission encountered in its discussions).

By the early 1990s, China's S&T system — and the broader society and economy — had changed dramatically from what it had been a decade before. S&T policy and reforms were important parts of this change. By promoting the importance of S&T knowledge for a modern society and the importance of people with such knowledge for the functioning of such a society and the organizations that make it up, the S&T-system reform process continued to call attention to the need for more economic and social reforms. In 1990, a new and ambitious vision for China's S&T future was introduced in the National Medium- and Long-term Science and Technology Development Program. The successful implementation of this plan presumes continuing institutional reform not only in the S&T system itself but also in the economy, in law, and in social-security arrangements. Followed as it was by the critically important Decision on Issues Concerning the Establishment of a Socialist Market Economy Structure, made at the Third Plenary Session of the 14th Party Conference in November 1993, the stage was set for important new initiatives in S&T and the economy as China began to prepare for the Ninth Five-Year Plan and the countdown to 2000.

Another National S&T Conference in 1995 and the May 1995 Decision on Accelerating Scientific and Technological Progress thus can be seen as occurring at the convergence of four factors:

- The experience of more than a decade of successes and failures with S&T reforms and with the other S&T programs alluded to above;

- Important streams of new reform thinking based on the idea of a socialist market economy, thinking that in turn is influenced by the reality that China's rapid economic growth has been driven more by economic activities outside the state sector than by those within it;

- The overall improvement of China's technological level during the reform period as a result of the very large expenditures on "technological renovation" (*jishu gaizao*) over the past 15 years (although some of this technology has come from China, a great deal of it has come from massive procurements of foreign technology; the introduction of advanced foreign technology continues with the surge of foreign investment — often in high-technology fields — that occurred in the early 1990s); and

- The impending competitive challenges China will face with its eventual admission into the WTO, and the fact that its domestic industries will face even more competition than they have thus far under the open-door policy of the past 15 years (to meet the challenges that WTO membership will bring, China is seeking to reorganize its industrial structure to ensure that it has firms ["pillar industries" or "national champions"] that have the size, technology, and managerial skill needed to compete in the global economy).

The convergence of these factors poses enormous challenges and opportunities for China's S&T policy and reforms. As a number of individuals mentioned to the Mission, the reforms since the 1985 Decision have had greater influence on the operating mechanisms of research institutions than on the broader problems of coordinating S&T activities for society as a whole. The very nature of the convergence of the four factors identified above means that a number of contradictory forces must be managed, and this fact, in turn, again calls attention to broader issues of the coordination and integration of an NSI.

Among the possible contradictions, the following stand out:

- State support for the creation of pillar industries must be reconciled with policies for an economically liberalized domestic market and with a liberalized foreign-trade and foreign-investment regime. Domestic R&D polices will have to be synchronized with the technology flows found in the global economy.

- R&D and overall technology management in pillar industries must come to resemble the technology strategies of the modern multinational corporation, not those of ministerially guided state enterprises of the past. S&T policy, therefore, must be developed in the context of changing industrial organization, the contours of which have yet to be fully understood.

— Because the S&T interests of the dynamic nonstate sector will not always converge with those of state-owned and state-directed pillar industries, ways must be found to reconcile the interests of both in a national S&T policy.

The potential for government policies to be internally inconsistent and for government agencies to work at cross-purposes in the face of such challenges is quite high. China, of course, is not alone in this: many other nations have the same problems as they attempt to adjust to global forces. What sets China apart is that it is in the throes of attempting to adjust to and exploit the globalization of industry and technology while continuing the still incomplete process of reforming its old S&T system.

Role of ministries in S&T decision-making

Among the many interesting issues for the future reform of S&T in China is the changing role of the national production ministries that, in the past, had important influence over the direction of S&T activities.

The changing role of ministries in Chinese S&T and economic reforms can be thought of in two ways:

— Their role vis-à-vis the central government and its coordinating commissions (SSTC, State Economic and Trade Commission [SETC], State Planning Commission [SPC], and SEdC); or

— Their role vis-à-vis local government.

In the pre-reform economic system, industrial ministries played a large role in decision-making, including S&T decision-making. Plans for research and production were developed within the ministerial system, and most of the assets for research and production within a given industry were controlled by the ministry. Within the state sector today, this continues to be true. Ministries run institutions of higher education, for instance, somewhat independently of SEdC, although these remain under the policy guidance of SEdC. Similarly, SSTC can set policy guidance for the ministerial S&T system — on reform policy, for instance — but because policy is separate from budgeting, large areas of discretion remain with the ministries. Because SPC has a role in national budgeting and SETC has a role in dispensing technical renovation funds and in SOE-management reform, the power of these commissions over ministries is even greater.

When viewed in relation to local government, a somewhat different picture emerges. In contrast to the situation in the former Soviet Union, centralization within a ministerial system was considerably weaker in China as a result of conscious decentralization of decision-making going back to the late 1950s. As a result, local governments — with a local industrial bureau matching the national ministry — have had a significant role in economic decision-making. Thus, in considering the role of ministries in the past, it is helpful to think in terms of two systems of organization leading to a complex pattern of "dual leadership." Although there are changes in course, as noted below, these systems continue today.

The first system is a vertical one, led by a central ministry in Beijing. For large SOEs and many large and nationally important research institutes, the central ministry provides primary leadership. Crosscutting the vertical system, however, is a horizontal one led by local government. Within local government are industrial bureaus accountable both to local political-leadership bodies and to the central ministry. Many SOEs and research institutes are under the leadership of these bureaus while also following broad, industry-wide directives set by the ministry.

Broader economic and managerial reforms now in course have created much confusion about the role of the ministries. Some ministries (for example, Electronics, Shipbuilding) have already converted most of their assets into large corporations, and this may be the direction other ministries will go. However, it remains unclear what the residual role of government is to be. This was evident during our visits in Liaoning. With the growth of managerial autonomy and the processes of "corporatization" and "enterprization" of SOEs and institutes, the prerogatives of industrial bureaus have become an issue. In Liaoning, for instance, the local industrial bureaus still appoint the directors of the successfully reformed institutes we saw but do not interfere in most other aspects of operations. Because the local industrial bureaus no longer provide any budgetary support, the newly "enterprized" institutes expect to enjoy the autonomy of a company operating in a market environment. However, the local bureaus are still exacting a percentage of the institutes' revenues, as in the past. A clearer definition of the role of government in this case appears to be a pressing matter affecting the future development of the institutes involved.

Similarly, with the corporatization of large SOEs into limited-liability share-issuing entities, the question of ministerial control is again an issue. At Anshan, for instance, we were told that the Anshan Iron and Steel Works (Angang), now being reformed into a company, was already competing in the market against other steel works, such as Baoshan, that are also under the Ministry of the Metallurgical Industry. Angang's Iron and Steel Research Institute still has a close relationship with the S&T system of the ministry, implying that — as in the past — its work is intended to serve the whole industry over which the ministry has cognizance. Clearly, at some point, contradictions and conflicts of interest are likely to arise if the research results from Angang's laboratories are transferred through ministerial mechanisms to a competitor in the market, such as Baoshan. To further confuse the situation, Angang is in the process of negotiating foreign participation in the financing and management of one of its subsidiary enterprises, creating a new level of ambiguity of ownership within the structure of the overall complex.

Given complexities of this sort, the pace of reform in the industrial economy — especially reforms dealing with shareholding and property rights — seems to be becoming a limiting factor in the reform of the industrial S&T system and in the implementation of an effective national industrial S&T strategy of the sort that recent policy documents have called for.

Central-local relations

The Mission also came to believe that the changing nature of the relationships between the central and local levels of government in S&T will develop as an issue in future.

Central-local relations in China have always been very complex, and they appear to be becoming more so. As a result of decentralization decisions that go back to the 1950s, local governments (provinces, cities) have substantial powers. Nominally, China is a unitary state (not a federation) in which powers of local government are delegated from the centre. The organization of local government thus in many ways resembles that of the centre, with local commissions and industrial bureaus having roles analogous to central commissions and ministries. These local entities, however, are not simply branches of the central entities. They tend, instead, to be agencies of local government. Many of China's SOEs and much of the S&T system are under the local governments. Large and nationally important enterprises and research institutes, however, are centrally controlled. Nevertheless, these units are also subject to local influence and coordination.

In the reform era, the wealth and power of many local governments relative to the centre have increased, although the wealth and power among local governments have become considerably more unequal. With the funding changes that have reduced the vertical income of national institutes and institutions of higher education and encouraged the development of horizontal revenue flows, the possibility of local governments increasing their influence with national S&T units has increased.

The Mission had opportunities to observe central-local S&T relations in four different settings (excluding Beijing) and found that there were considerable variations. Indeed, we identified at least three different models at work.

The Guangdong model is at one extreme. Guangdong is a leader in reform and is extensively integrated into the world economy. Because Guandong is a very wealthy province — but one that traditionally had a much more limited concentration of S&T assets (numbers of scientists and engineers, institutes, and institutions of higher education) than other places visited by the Mission — the role played by the provincial government, in the direction of the S&T work done in the province, is very strong. Reportedly, 70% of the government-funded R&D in the province, including that done by nominally central institutes, is funded locally. The local branch of the CAS has long been closely linked with the Guangdong Academy of Sciences in ways that are not seen elsewhere (they share the same building and have overlapping staff, for instance).

A second model could be based on the Liaoning and Shaanxi experiences. Both provinces have had abundant S&T assets, especially those controlled by the centre. In addition to having S&T assets, they have been centres of strategic and domestic high-technology industries. However, both provinces have been slow in reform and are less well integrated into the international economy. The provincial governments seem to have less money to induce cooperation from central units. The Shenyang branch of CAS, for instance, seems oriented much more toward the central CAS than toward provincial concerns.

A third model is found in the city of Shanghai. Like Liaoning and Shaanxi, Shanghai has abundant S&T assets, including those under central control and those under the Shanghai government. Although Shanghai was slow in reform at first, it is now moving rapidly. Shanghai and the lower Yangtze Valley are becoming favoured destinations for foreign investors in technology-intensive industries. Shanghai has long had a problematic relationship with the centre over centrally controlled SOEs and institutes whose location in Shanghai put additional financial burdens on the city without providing appropriate returns. However, as Shanghai's wealth continues to grow and as the city grows into its role as one of the most important — if not the most important — point of connection with the global high-technology economy, it seems inevitable that central institutes will become integrated into local and regional initiatives. An interesting case is the creation of the new Shanghai Academy of Sciences. Although this academy, at the moment, consists only of institutes that belong to the city, it may provide an organizational structure into which central institutes might fit.

Chapter 5

INNOVATION
AND A NATIONAL SYSTEM
OF INNOVATION

OECD (1994, p. 3) provides a useful definition of *innovation* as "the transformation of an idea into a new or improved product introduced on the market or a new or improved operational process used in industry and commerce or into a new approach to a social service."

This description brings out the point that technological innovation involves more than R&D — it also involves the workings of the marketplace. Innovation can, of course, occur in any human activity, although it is primarily thought of in the context of industrial production.

The overriding objective of China's S&T policy, as stated in the May 1995 Decision, has been to implement the idea that S&T is a primary productive force in all fields. In terms used in industrialized countries, this policy measure is one that calls for the promotion of technological innovation throughout society and the economy. In a country with such a policy orientation, it is now considered useful to seek to identify the elements of an NSI and then to use this concept as a basis for policy formulation.

The Mission found this a helpful approach to understanding the complexity of the Chinese S&T system and to identifying the roles played by the key stakeholders. It also helped us identify any gaps within the system.

Concept of a national system of innovation and its use as a policy framework

There have been many attempts in industrialized countries to concisely define an NSI. OECD (1994, p. 3) calls such a system "a network of institutions in the public and private sectors whose activities and actions initiate, import, modify and diffuse new technologies." An alternative, somewhat fuller definition (OECD 1994, p. 3) is

> a system of interacting private and public firms (either large or small), universities and government agencies aiming at the production of science and technology within national borders. Interaction among these units may be technical, commercial, legal, social and financial, inasmuch as the goal of the interaction is the development, protection, financing or regulation of new science and technology.

An NSI can be thought of as a set of functioning institutions, organizations, and policies that interact constructively in the pursuit of a common set of social and economic goals and objectives and that use the introduction of innovations as the key promoter of change.

The four key interests, then, of any country can be thought of as

– Ensuring that it has in place a set of institutions, organizations, and policies that give effect to the various functions of an NSI;

– Ensuring that there is a constructive set of interactions among those institutions, organizations, and policies;

– Ensuring that there is in place an agreed-upon set of goals and objectives consonant with an articulated vision of the future that is sought; and

– Ensuring that there is in place a policy environment designed to promote innovation.

Table 2 lists what we consider to be the essential functions of any NSI. In the next section, we proceed to look at the roles played by a variety of Chinese institutions in those functions.

Stakeholders in China's NSI

In industrialized countries, *stakeholder* is used in discussions of systems of many kinds, including discussions of NSIs, to indicate the institutions and individuals who are participants in the system in question or whose activities are significantly affected by the operation of that system. It is necessary to include in any analysis of an NSI a clear listing of the relevant

Table 2. Functions of an NSI.

General functions	Specific functions
Core functions of government	
Policy formulation and resource allocation	– Monitoring, review, and formulation of policies and, in some countries, plans concerning national S&T activities – Linkage to other policy domains (such as the economy, trade, education, health, environment, and defence) – Allocation of resources to S&T from overall budgets and first-order allocation among activities – Creation of incentive schemes to stimulate innovation and other technical activities – Provision of a capacity for implementing policies and coordinating appropriate activities – Provision of a capacity for forecasting and assessing the likely directions of technical change
Regulatory	– Creation of a national system for metrology, standardization, and calibration – Creation of a national system for the identification and protection of intellectual property – Creation of national systems for the protection of safety, health, and the environment
Implementation functions	
Financing	– Management of financing systems appropriate to the implementation of the other functions of the system – Use of government's purchasing power as a stimulus to innovation in the production of the goods and services it requires
Performance	– Execution of scientific or technological programs, including research of all kinds and technological development – Provision of scientific services – Provision of mechanisms to link R&D outputs to practical use – Provision of linkages to regional and international S&T activities – Provision of mechanisms for evaluating, acquiring, and diffusing best-practice technologies – Creation of innovative goods, processes, and services embodying the results of S&T activities
HRD and capacity-building	– Provision of programs and facilities for the education and training of S&T personnel – Creation of institutional capacity in S&T – Provision of mechanisms to maintain the vitality of the national S&T community – Stimulation of public interest in and support of national initiatives in S&T
Infrastructure	– EOM of information services (including libraries, databases, statistical services, a system of indicators, and communications systems) – EOM of technical services (such as metrology, standardization, and calibration) – EOM of a system of awarding, recording, and protecting intellectual property – EOM of mechanisms to ensure the protection of safety, health, and the environment – EOM of major national facilities for research

Note: These functions — both policy-related and implementation-related — are carried out by different stakeholders in any country's national system of innovation, with the particular combination being unique to that country. EOM, establishment, operation, and maintenance; HRD, human-resources development; NSI, national system of innovation; R&D, research and development; S&T, science and technology.

stakeholders, and this we will now do. (Given that the Mission spent only 3 weeks in China, it is likely that we have failed to identify some stakeholder groups, so the listing below may well be incomplete.)

Policy-making institutions

A number of important national commissions, ministries, and institutes have significant roles in China's NSI, including the following:

- SSTC, with its important roles in policy development and program implementation;
- SPC, which is directly involved in financing a number of important S&T programs;
- SEdC, which has particular responsibilities for the activities of institutions of higher education;
- SETC, which has an important role in the technological renovation of enterprises;
- The State Commission for Restructuring the Economic System (SCRES), whose general economic reforms have in many ways interacted with reforms in the S&T sector;
- The many sectoral ministries (responsible for individual sectors of the industrial economy), some of which are now undergoing transformation into what look like holding companies within the socialist market system; and
- Some policy research institutes, particularly the National Research Centre for Science and Technology for Development (NRCSTD) and ISPM–CAS.

Similar organizations at the provincial and municipal levels also have important roles.

Principal S&T institutions

The following six kinds of organizations constitute the heart of the Chinese NSI:

- Research institutes;
- SOEs;
- Private, joint-venture, and urban collective enterprises;
- Universities;

Table 3. Estimates of the number of R&D institutions in China.

	1987	1988	1989	1990	1991
R&D institutions under government departments	7 292	8 169	8 456	8 576	8 188
Natural S&T	5 222	4 933	5 011	5 084	5 127
Social sciences and humanities	346	342	343	332	336
Information and documentation		396	405	414	416
Other fields, including health	1 724	2 498	2 697	2 746	2 309
Institutions under the State Council	1 033	1 000	1 010	1 027	1 035
Institutions under local governments	4 189	3 933	4 001	4 057	4 092
Technological-development institutions under large- and medium-scale enterprises	5 021	5 525	7 215	8 116	8 792
R&D institutions at institutions of higher learning	1 514	1 715	1 739	1 666	1 676
Collective and private technological-development institutions	2 013	4 870	6 424	8 523	
Collective	1 536	3 407	4 359	6 047	
Private	422	1 239	1 740	2 243	
Status unknown	55	224	325	133	
Total calculated by Yang[a]	16 477	20 279	23 838	26 881	
Total	21 062	25 212	28 845	31 965	

Source: Yang (1994).
Note: R&D, research and development; S&T, science and technology.
[a]There is an unexplained and large discrepancy in the totals calculated by Yang.

— Defence research institutes and enterprises; and

— TVEs.

All of these organizations have been significantly affected by the reforms of the last decade. Estimates of the number of such organizations vary, but one attempt to estimate the numbers of those heavily engaged in R&D — and therefore having a significant potential to promote innovation — is presented in Table 3.

New organizational forms created by the reform process

The reform of R&D institutes in China can be seen as having had a series of objectives:

1. To reduce dramatically the dependence of many institutes on annual budget appropriations from the state;

2. To introduce externally reviewed or market-driven competitive processes for funding research; and

3. To establish, within existing institutes, specialized groupings designed either to concentrate high-level teams on programs of strategic research or to serve as improved vehicles for the commercialization of technologies.

The principal organizational structures created to accomplish these objectives, and now important stakeholders in the NSI, are the following:

- NNSF;

- National key laboratories, selected on the advice of NNSF and funded by SSTC, SPC, and, in some cases, the World Bank;

- The research laboratories of SEdC;

- The ERCs supported by SSTC; and

- The ERCs supported by SEdC (and some about to be created with World Bank assistance).

NNSF is a completely new and independent organization; the other organizations are special components of existing institutes or universities.

In addition, SSTC has recently begun to plan a new set of at least 10 national research centres (to be established under the Ninth Five-Year Plan) that appear to come in at least three different organizational forms (described in Chapter 3):

- The geographically distributed network;

- The consortium in a common location; and

- The new centre in new facilities.

Organizations of the scientific community

CAST and its constituent societies are stakeholders in China's NSI as well.

Relevant financial institutions

Financial institutions that are major stakeholders in the NSI include the following:

- Banks that give loans for S&T and innovation-related activities; and

- Venture-capital organizations (at least one of which, in Guangzhou, is "owned" by a provincial STC).

Regulatory bodies

Regulatory bodies that are stakeholders in the NSI are of the following types:

- Organizations responsible for the protection of intellectual property;
- Organizations for the protection of health, safety, and the environment; and
- Organizations concerned with standards, calibration, and metrology.

Of the financial institutions and regulatory bodies in China, the only one interviewed by the mission — and that only briefly — was a municipal EPA in Beijing.

Other stakeholders

There are a number of foreign companies, foreign (development-assistance) agencies, and multilateral bodies active in China's NSI, and their activities need to be taken into account.

Roles of stakeholders in the functions of China's NSI

In Tables 4–8, we explore the roles of different stakeholders in China's NSI to the extent that we are able to do so given our brief visit to the country. We use a series of tables (matrices) to map out the different functions of the NSI and demonstrate that this mode of analysis can be performed at different levels of generality (such as at the level of the whole system), for some subset of issues (such as financing issues), or for some subset of stakeholders (such as government institutions).

Table 4 explores the roles of government stakeholders in the policy and regulatory functions of the Chinese NSI, and Tables 5 and 6 explore different stakeholders' relationships to the financing functions of the system. Table 7 explores the roles of a broader list of stakeholders in the implementation functions. Table 8 also uses the technique of displaying information in matrix form to show the participation of stakeholders in the programs administered by SSTC.

One of the important considerations that affects the functioning of an NSI is the extent to which the relevant governments can bring about an integration of their policies and funding programs to produce a positive policy environment that encourages entrepreneurial activity and technological innovation. The OECD countries, in a series of publications emanating from a major program of research on technology and economic policy (e.g., OECD 1991, 1992), concluded that industrialized countries

Table 4. Roles of government stakeholders in the policy and regulatory functions of the Chinese NSI.

Stakeholders	Policy and resource-allocation functions						Regulatory functions		
	Policy setting	Policy integration	Resource allocation	Incentives for R&D	Policy and program coordination	Technology foresight	Health, safety, and environment	Intellectual property	Metrology, standardization, and calibration
State Council	•	•					•		•
SSTC	•	•	•	•	•	•		•	•
SPC	•	•	•	•	•	•			•
SEC	•	•	•	•					•
SEdC	•	•	•						
Sectoral ministries		•							
CAS			•	•					
CAAS									
Provincial commissions	•	•	•	•			•		•
Municipal commissions	•	•	•	•			•		
NRCSTD						•			
ISPM-CAS									
CAST									

Note: CAAS, Chinese Academy of Agricultural Sciences; CAS, Chinese Academy of Sciences; CAST, Chinese Association for Science and Technology; ISPM, Institute of Science Policy and Management [CAS]; NRCSTD, National Research Centre for Science and Technology for Development; NSI, national system of innovation; R&D, research and development; SEC, State Economic Commission; SEdC, State Education Commission; SPC, State Planning Commission; SSTC, State Science and Technology Commission of China.

Table 5. Government stakeholders and the financing functions of the Chinese NSI.

		Financing functions				
Stakeholders	Government budgets	Grants	Loans	Contracts	Tax incentives	Government purchasing
State Council	Approves budgets	Not involved	Not involved	Not involved	Has agreed to a variety of investment incentives	Would need to approve policy
SSTC	Advises on size and distribution	Provides grants		Provides contracts	Provides some incentives via Torch Program	
SPC	Provides budgets	Provides capital grants to ERCs via a World Bank loan		Provides contracts		
SEC		Provides grants		Provides contracts		
SEdC		Supports ERCs		Provides some contracts		
Sectoral ministries		Support some programs		Provide some contracts		
Provincial governments	Only some can provide budgets to their institutes	Provide some grants	May guarantee loans to enterprises in high-tech development zones	Some provide contracts		
Municipal governments	Only some can provide budgets to their institutes		May guarantee loans to enterprises in high-tech development zones	Some provide contracts		
Banks			Now major suppliers of loan financing			
NNSF		Major source of support, particularly to basic research		Provides contracts		

Note: ERC, engineering research centre; NNSF, National Natural Science Foundation; NSI, national system of innovation; SEC, State Economic Commission; SEdC, State Education Commission; SPC, State Planning Commission; SSTC, State Science and Technology Commission of China.

Table 6. Other stakeholders and the financing functions of the Chinese NSI.

Stakeholders	Financing functions (policy instruments)		
	Government budgets	Grants	Loans
R&D institutions			
Academies	Major recipients	Important recipients	
Government institutes (state, provincial, and municipal)	Some get government support; others are cut off	National institutes may get some; others are less likely to	Possible under some circumstances
Universities	Government grants cover most of staff salaries	A major source of R&D support	
Enterprises			
SOEs	Many need recurring subsidies		Some may get subsidized loans
TVEs	Not eligible		Interest rates may be subsidized
Spin-offs and NTEs	Not eligible	Can, in theory, compete; a few are successful	Eligible to have interest paid by government if located in a high-tech development zone
Joint ventures	Not eligible		Eligibility not clear

Stakeholders	Financing functions (policy instruments)		
	Contracts	Tax incentives	Self-financing schemes
R&D institutions			
Academies	Important recipients	May be eligible in some circumstances	Revenues from some spin-off companies
Government institutes (state, provincial, and municipal)	Important source of revenue; some depend entirely on contracts	Not clear under what conditions they are eligible to receive incentives	Some get into production activities to generate revenue for survival
Universities	A second important source of R&D support		Revenues from some spin-off companies
Enterprises			
SOEs	Some award contracts for R&D	New facilities in designated zones are eligible	Profitable ones are self-financing
TVEs			Should become self-financing
Spin-offs and NTEs		Eligible for most-favoured tax treatment in designated high-tech development zones	Should become self-financing
Joint ventures		Eligible for investment incentives	Have to be self-financing

Note: NSI, national system of innovation; NTE, new technology enterprise; R&D, research and development; SOE, state-owned enterprise; TVE, township and village enterprise.

Table 7. Stakeholders and the implementation functions of the Chinese NSI.

Stakeholders	Performance			HRD and capacity-building		Infrastructure		
	R&D innovation	Linkages and networks	Transfer and adoption including assimilation of foreign technology	Tertiary education and training	Institutional capacity creation	Regulatory (health, safety, and environment)	S&T information	Intellectual property
SSTC	Has extensive programs	Gives preference, in national programs, to proposals from more than one group	Supports ERCs and has other extensive programs		•	•	•	•
SPC	•	•	Supports ERCs	•	•	•	•	
SEC	•	•		•	•			
SEdC	•	•	Supports ERCs	•	•		•	
Sectoral ministries	•	•		•	•	•	•	•
NNSF	•	•		•	•		•	
Academies	•	•	•	•	•		•	
Government institutes (state, provincial, and municipal)	•			•	•		•	
Universities	Much activity concentrated in a few leading institutions	A few involved	•	•	•			
SOEs	Performance varies from sophisticated to nil		•	•				
TVEs			Some are major recipients					
Spin-offs and NTEs	Many have activities in R&D	Some may participate	•					
Joint ventures	Some involved		Many involved in import of technology					

Notes: ERC, engineering research centre; HRD, human-resources development; NNSF, National Natural Science Foundation; NSI, national system of innovation; NTE, new technology enterprise; R&D, research and development; S&T, science and technology; SEC, State Economic Commission; SEdC, State Education Commission; SOE, state-owned enterprise; SPC, State Planning Commission; SSTC, State Science and Technology Commission of China; TVE, township and village enterprise.

Table 8. Role of SSTC and its major programs in supporting the S&T activities of the principal stakeholders in China's NSI.

Stakeholders	Key Technology (policy and funding)	Key Laboratory (implementation, with SPC and SEdC)	Engineering Research Centres	Key Industrial Projects	Industrial Technology Extension	863 (funding and policy)	Spark (funding and policy)	Climbing (funding and policy)	Torch (funding and policy)	National S&T Achievements (funding and policy)	Trial Production and Appraisal	Agenda 21 (policy)
SSTC institutes	•	•	•			•	•	•	•	•		•
CAS	•	•	•	•		•	•	•	•	•	•	•
CAAS	•	•	•			•	•	•	•	•		•
Sectoral ministry institutes	•	•	•	•	•	•	•	•	•	•	•	•
Provincial STC institutes	A few									•		
Municipal STC institutes	A few									•		
Universities	•	•	•	•	•	•	•	•	•	•	•	•
SOEs	•			•	•	•			•	•		
Spin-offs	•								•	•	•	•
NTEs	•				•				•			
TVEs					•		•		•	•	•	•
Joint ventures	•											

Note: CAAS, Chinese Academy of Agricultural Sciences; CAS, Chinese Academy of Sciences; NSI, national system of innovation; NTE, new technology enterprise; S&T, science and technology; SEdC, State Education Commission; SOE, state-owned enterprise; SPC, State Planning Commission; SSTC, State Science and Technology Commission of China; STC, Science and Technology Commission; TVE, township and village enterprise.

need to improve the integration of their various programs and policies. This approach is easier to prescribe than to implement, and each country has to devise a system suited to its own political culture.

In China, we have seen that at least four important commissions are active in policy formulation and program delivery in ways designed to promote S&T reform — and hence to stimulate innovation in the emerging socialist market economy — but we have been unable to gain an insight into how their various activities are coordinated. The commissions involved are SSTC, SPC, SEdC, and SETC. We do not know of any mechanism currently in place to facilitate a high-level interaction among senior officials of these bodies with a view to harmonizing their interventions. Such interactions may already occur on a case-by-case basis, but we have no indication of their effectiveness. Even beyond this set of commissions, there are other ministries that have interests in, and policies for, S&T, but we have heard of no forum in which these ministries could participate in a full discussion of their interests in S&T. Furthermore, China has S&T policy activities going on at the provincial and municipal levels, so the challenge of seeking some harmonization of all of the separate policies and programs is daunting.

Chapter 6

POLICIES FOR BASIC RESEARCH IN CHINA

Current policy in China

From as early as the 1978 National Conference on Science and Technology, the Government of China has been committed to a policy under which basic research should be permitted to grow gradually on a stable basis. The most recent, authoritative statement of Chinese policy for the support of basic research is to be found in the May 1995 Decision, Section VI — "Firmly Tighten Basic Research" (*National Affairs* 1995). That section describes "the missions of basic research" as being

> to explore natural laws, strive for new discoveries and inventions, accumulate scientific knowledge, establish new theories, and provide the new theories and methods for rebuilding the world. ... to give priority to state-set objectives and consider it its central task to provide the power for national economic and social development. ... to proceed according to the principle of "catching up" in some but not all areas.

That section also argues that basic research should "stress key issues, pooling its resources on tackling those projects that will likely play a significant role in promoting national economic and social development." This could serve equally well as a statement of what industrialized countries currently think of as strategic or even precompetitive research — that is, fundamental research at the frontiers of knowledge, undertaken in areas of science considered as being of long-term significance for future technological development. We will return later in this chapter to some of the implications of this mission statement for basic research in China, as seen from the vantage point of the industrialized countries.

The Decision makes a commitment "to continue to increase investments in 'basic research' ... [so that] funds for 'basic research' should gradually account for a larger percentage of research and development budgets." (SSTC would like the share of GERD allocated to basic research to grow from the current 6 or 7% to about 10% by 2000, by which time GERD is supposed to increase from 0.7% to 1.5% of GDP.) The Decision goes on to argue that basic research should be "organically merged with the training of proficient personnel."

One unique cultural feature of Chinese policy that has important effects in the area of basic research is the use of eight-character phrases to convey the essence of policies. This long tradition has become a source of pithy slogans that institutions and individuals are expected to interpret and implement. In recent years, three such slogans have been coined to capture important policy initiatives; they are, together with the Mission's own informal translations, the following:

- *Wenzhu yitou, fangkai yipian*, "anchor one end securely, let the other roam free";

- *Ding tian li di, hou lai ju shang*, "go forward with your head in the clouds, your feet on the ground, start late, but end up ahead"; and

- *Mian xiang, yi kao pan gao feng*, "facing the economy, rely on S&T to climb high mountains."

The first slogan applies to basic research and is interpreted in different places as meaning that the government needs to secure the funding of certain important areas — including basic research — that are in the public interest, but for which there is little market support; and, at a different level, institutes should support the best researchers with government funding and let the others look in the marketplace for support. The second slogan refers to policies for the support of high technology and can be interpreted as calling for a strategic view of research — it should be innovative and at the leading edge of science but in areas likely to have practical applications. The third slogan is an expression of Chinese determination to mobilize S&T to achieve great economic success.

Chinese activities in basic research have been substantially affected by a series of significant reforms in the last decade, of which the most important was arguably the creation of the NNSF in 1986. The founding of NNSF formalized the introduction of peer-reviewed, competitive project-grant

Table 9. NNSF grants to research projects.

	Major projects	Key projects	General projects
Participating researchers	20 or 30 minimum; up to 100 in some cases	About 20 per group	Small groups or individuals
Duration of grant (years)	5	4 or 5	3
Share of NNSF budget (%)	About 20		About 70
Description of activity	Multidisciplinary; related to an economic or social goal	Work at the frontiers of a discipline	Researcher determined
Designation of priorities	Done periodically by a large array of expert panels	Related to areas of existing Chinese strength and to needs of the high-tech industry	Researcher determined

Note: NNSF, National Natural Science Foundation.

funding in Chinese basic research, replacing the former reliance on annual institutional budgets as the primary source of funding. In so doing, China has adopted an approach that is common throughout the industrialized world and that is believed in industrialized countries to be an effective way of stimulating scientific activity of high quality. Industrialized countries are continuing to experiment with ways to inject into the peer-review process the concept of relevance and perceptions of the future importance of emerging fields of activity as criteria.

NNSF had a budget of more than 1.5 billion CNY during the Eighth Five-Year Plan (1990–95), and its annual expenditures have been growing faster than inflation in recent years. To allocate these resources, NNSF has put in place a set of funding programs designed to balance the objectives of providing the scientific underpinnings of national programs of social or economic development and of providing opportunities to individuals or small groups to pursue more speculative ideas. At present the foundation allocates the bulk of its funding to three categories of grants (Table 9).

Competition for NNSF grants is fierce, with only about one application in seven being successful. About two-thirds of NNSF funds go to university-based research groups, with the remainder going primarily to researchers from CAS.

Since early in this decade NNSF has administered a fund (most recently, 4 million USD) contributed by the Ford Motor Company of the United States to support research in areas of interest to the future development of the automobile industry — an example of a multinational corporation financing strategic research in a developing country that has a combination of real research strengths and low costs of doing research.

In Guangdong Province, the Mission learned that a provincial foundation established in 1987 was currently allocated an annual budget of about 10 million CNY.

In addition to creating and financing NNSF, the government had earlier, in 1984, begun to designate and finance a series of state key laboratories — located primarily within a small group of universities and institutes of CAS — within which modern equipment for research would be installed and operated. Furthermore, the government went on to designate a significant number of open laboratories, which were encouraged to promote cooperative activities in basic research involving domestic and international scientists.

During the Eighth Five-Year Plan, SSTC introduced the National Basic Research Priorities Program (the Climbing Program), the goals of which were to

- Gradually increase the (financial resource) input into basic research;

- Support basic research through diversification of funding channels;

- Train researchers, improve their quality, and encourage them to work at the world frontiers of knowledge;

- Support open laboratories, improve working conditions, and generate a better academic environment; and

- Promote international cooperation and exchange.

A more recent initiative of SSTC is the creation of a small series of national research centres (possibly as few as 10) designed to combine the strengths of varying numbers of institutions of different kinds in both basic and applied research. These centres will come into operation during the Ninth Five-Year Plan (1996–2000).

Observations from the Mission

During our various meetings we received a variety of comments regarding the performance of basic research in China that give a "flavour" of the opinions of active scientists:

- Many scientists appear to believe that individual research grants are getting smaller and are for shorter duration [but could this reflect, in the minds of some, a nostalgia for the less competitive system of the past?] and that there is

    - A shortage of funds, especially for long-term research,

    - Limited room for choice of research topic,

- A lack of nongovernmental sponsorship for basic research, and
- A lack of rewards for groups that make contributions leading to widespread social benefit;
— Other scientists argue that more able scientists today can get better funding for their work from NNSF than was available 10 years ago whereas less able scientists find conditions more difficult;
— NNSF believes that there is a need to
- Improve the budget allocated for updating equipment and purchasing scientific literature,
- Continue to stress the need for young scientists to get into basic research,[7] and
- Get more enterprises to support and engage in basic research;
— A number of people interviewed believed that SSTC should devote more attention to the problems of research in the universities; and
— There is widespread support for the idea that scientific research in the universities should be closely linked to the training function performed by those universities.

Also, according to SEdC data,
— 60% of all basic research is done in universities that receive only 4.7% of total GERD;
— 22% is done in the more than 5 000 research institutes belonging to the different levels of government; and
— 8% is done in enterprises (primarily in spin-off enterprises that were formerly parts of a research institution, rather than in traditional SOEs).

The huge number of institutions in China today is a source of concern to policymakers, and we in the Mission can identify one dilemma that relates to the role of the university system in research, particularly in basic research. According to data we received, China today has 1 058 universities, of which

— Fewer than 100 can claim to be research intensive;
— About 200 have the right to grant doctoral degrees in at least some departments; and

---

[7] According to NNSF data, in 1986 only 1.5% of project leaders were under the age of 35; in 1995, this had risen to 29%.

— About 400 have the right to grant master's degrees, again in at least some departments.

According to data cited to the Mission by SEdC,

— The top 50% of universities receive about 65% of all government S&T funding available to the universities as a group; and

— The top 100 receive about 80% of those funds.

The concentration of nationally identified facilities is also very marked. From information provided to the Mission, we calculated that there are 201 state key laboratories, SEdC laboratories, and ERCs operating in China today; 100 of them are located in the top 11 universities (20 of them in Tsinghua alone). This already represents a very high degree of concentration, and given that these special facilities are designed to attract able scientists and be well equipped (at least in relative terms), we would expect that scientists from these universities would receive a significant share of the funding awarded by competitive grant by the NNSF.[8] This leads us to question how much research content there can be in many of the universities that currently grant higher degrees but are not on the list of research-intensive institutions. There is apparently a program (the 211 Program) in place to attempt to concentrate R&D activities in the top 100 universities, but the Mission was given no information on that initiative.

Some people we spoke to appeared to interpret the Chinese policy for basic research as meaning that such research should be exclusively the activity of the universities and a few CAS institutes. We do not agree with that interpretation. We believe neither that basic research should be confined mainly to the universities nor the converse — that the universities should be involved solely in basic research. As we will discuss below, there are new forms of arrangement for strategic research that seem to us highly appropriate for China, in which a whole range of institutions can and should participate, and in which there is ample role for the effective combination of research and the training of new researchers.

We have sensed that there is considerable debate surrounding the role of CAS within Chinese science, including within basic research. People within the organization argue that basic research is "having a difficult time" within the Academy and that younger scientists are moving toward the greater financial returns available in the marketplace. (However, we must ask how this observation can be reconciled with NNSF's contention that the proportion of team leaders who are under the age of 35 has risen

---

[8] This statement represents a judgment on our part; we have not seen data on the apportionment of NNSF grants to recipient institutions.

substantially in recent years.) Outside the Academy there are those who wonder about its place in the new China. Some ask whether the long-term research conducted in CAS might not equally be carried out in a university setting, which would render it more accessible for training graduate students.

CAS plays an important role in Chinese big science (see Chapter 2), and we were told that during the Seventh Five-Year Plan, China constructed 10 major big-science facilities:

– An electron–positron collider in Beijing;

– A heavy-ion accelerator for Lanzhou;

– A synchrotron-radiation laboratory in Hefei;

– A 2.16-m optical telescope;

– The Beijing solar magnetic fields telescope;

– The Beijing remote-sensing satellite ground station;

– The 5-MW low-temperature nuclear heating tandem accelerator (Beijing, HL-13);

– A long-wave time system; and

– The China HL-1 Tokomak.

For the Ninth Five-Year Plan, the state has approved an allocation of 2 billion CNY for a series of projects:

– A particle-beam factory (type not yet decided);

– A nuclear-fusion experimental facility;

– A large astronomical telescope;

– A major facility for research in the life sciences; and

– A monitoring and surveying network to support ocean, land, and space engineering.

However, we obtained no information on the analytical processes that led to this significant allocation of funds to big science and were given no information on the relative emphasis that China is likely to place in the years ahead on investments in national big-science facilities in comparison with financing Chinese participation in international projects.

Some reflections on international experience and practice

In a variety of industrialized countries, the use in policy discussions of the phrase "basic research" is in decline. In Canada, for example, the national statistical organization has abandoned the attempt to accumulate data under the heading of basic research, on grounds that deciding what constitutes basic research was too subjective. This, however, does not in any sense represent a diminution of policy interest in supporting long-term research that seeks a deep understanding of natural phenomena; it is simply a recognition that policy now addresses issues other than those that in the past were raised during analysis of different disciplines. In the industrialized countries, basic research has come to embody a whole range of activities (see the full descriptions in Chapter 2):

- Curiosity-oriented research;
- Strategic research; and
- Big science.

Closely related to the concept of strategic research is that of precompetitive research, in which enterprises that are competitors in the market join together — often in research consortia — to undertake long-term research projects that are both costly and technologically risky but are in areas of potential significance to their industries.

We have heard much discussion in China about the importance of basic research, but there always seems to us to be great vagueness about what actually is being discussed. We find the term *applied basic research* particularly confusing. We can, however, see good justification for China remaining active in curiosity-oriented research, strategic research, and big science.

One feature of current policy thinking in industrialized countries is the attention paid to understanding how research works, in addition to giving consideration to what the objectives of research should be. The social organization of scientific research, particularly of research that could be designated as basic, as practiced in the industrialized countries differs from that of technological development in several important ways, summarized in Table 10. These different features influence the ways in which these activities are organized.

The dominant features of the structure of contemporary scientific activity in the industrialized countries were described in the recent work of Gibbons et al. (1994), who argued that

- Knowledge is ever more produced in the context of its applications, and there are greater expectations that support of research will lead

Table 10. Social organization of scientific research and technological development in industrialized countries.

Characteristic	Scientific research	Technological development
Motivation	Intellectual: general curiosity about natural phenomena	Economic or social: desire to solve specific problems
Attitude to information	Operates on basis of wide dissemination of ideas and information to permit their validation (the "open scientific literature" is a key means of dissemination)	Often operates on the concept of ideas as proprietary knowledge; will tend to use instruments of intellectual property as vehicles for disclosure
Usual location of activity	Universities and government laboratories; a few large industrial labs	Primarily in industrial laboratories in the private sector (difficulties are experienced when linkages to the user are weak)
Participants	A high proportion have had doctoral-level training in a specific discipline	Greater emphasis on practical experience beyond academic training
Structure of groups	Traditionally, in many disciplines, this is the domain of the individual researcher, but this model is rapidly disappearing as a result of factors related to the increasing cost of facilities and the increasing potential for (electronic) networking	More likely to contain an interdisciplinary mix of scientists and engineers; in some cases, there may even be economists or other social scientists as well

directly to economic and social benefits for the nation providing the support;

– There is an inescapable trend toward larger and more interdisciplinary teams working in more transdisciplinary research activities;

– There is a growing diversity of participating organizations to be found in today's research teams (there can be a blurring of project or program); and

– There is a continuing trend toward greater international linkages within research teams.

Such an attempt to explain a complex new mode of behaviour comes with its own built-in contradictions. Debates can develop about the relative priority to be given to national versus enterprise interests or about the nature of national benefits to be derived from an international program.

We have seen some evidence that the new organizational forms being introduced in China are designed to respond to at least three of these trends, with the exception being the relative lack of evidence of emphasis in China on transdisciplinary organizations. Many newer bodies — including SSTC's national research centres (to be introduced during the Ninth Five-Year Plan) and the state key laboratories — still seem to be organized along traditional disciplinary lines, and this, we believe, is a shortcoming.

We would pose a number of questions about the management challenges facing some of the proposed new national research centres and will use the Shanghai Research Centre for Applied Physics (SRCAP) as an example to illustrate our point (Box 3). All of the participating institutions already have access to multiple channels of funding for their existing research activities and have ongoing projects in their fields of expertise. Will the director of SRCAP have under his or her direct control enough additional funds to use as a lever to build some coherence into the research programs of the participating groups? Will the participants contribute some of their existing funding to enable them to establish larger programs defined by the new centre and its advisory board? Will any institution joining the consortium make an annual contribution to the research budget of the new centre?

We ask these questions because we have heard that one unintended consequence of the reforms in research funding, at least in some institutes, has been to deprive the institute director of most of his or her power to influence the direction and coherence of the institute's programs. This power, it is argued, is effectively now in the hands of the principal researchers using the external grants that now finance the institute's programs. The potential in consortia such as SRCAP for such an outcome is likely to be large, unless appropriate steps are taken to give management real responsibility for the centre's work.

Box 3

Grouping of researchers by the Shanghai Research Centre for Applied Physics

The new SRCAP will group researchers from

- Eight state key laboratories (five are university based, and three are in CAS institutes);
- Six university laboratories (from three different universities);
- One joint laboratory (operated by the CAS and a fourth university); and
- Six other CAS institutes.

In industrialized countries where such centres are designed to undertake strategic research, great care is taken to ensure the extensive participation of representatives of enterprises from the industrial sectors expected to benefit from the research. A significant share of seats on a centre's board of directors would be filled by people drawn from enterprises, and most centres would have scientists from enterprises active in their research activities. The Mission has not learned whether the new Chinese national research centres will follow this pattern.

Chapter 7

THE HIGH-TECHNOLOGY SECTOR

Background to the reform process

China has had a commitment to new science-based technologies since the 1950s, but these had a chequered history through the period of intense development of military research and the disruption of the Cultural Revolution of the late 1960s and early 1970s. The reform process that commenced in 1978 has been directed to restoring that early commitment to the role of the science-based technologies as drivers of the civilian economy.

The National Conference on Science and Technology of 1978 identified 108 key projects in eight priority areas: agriculture, energy, materials, microelectronics, lasers, space, high-energy physics, and genetic engineering. This ambitious program was driven by the view that basic research was the wellspring for innovation (as in the old linear model formerly espoused in industrialized countries) and by the need to rehabilitate scientists and engineers and the R&D system after the Cultural Revolution. New institutes were created, leading to an increased expenditure on S&T in the government budget (4.8%, 4.9%, and 5.3% of total government expenditure and 1.5%, 1.6%, and 1.5% of gross national product [GNP] in 1978, 1979, and 1980, respectively).

It rapidly became clear that this approach was not producing in terms of short-term economic growth and that a more pragmatic approach to linking S&T to the economy was needed. The National Program for

Tackling Key Technology Problems of the Sixth Five-Year Plan (1980–85) identified relevant areas, of which only agriculture remained as a key area, and highlighted another four:

– Energy-resource development and energy concentration;

– The raw materials industries and geological exploration;

– Mechanical and electrical equipment and transportation; and

– New technologies and social development.

Support for high technology continued into the Seventh Five-Year Plan, adopted in 1981, and continued into the Eighth Five-Year Plan, with a growing emphasis on reforms leading to commercialization of technological activities and to new modes of cooperation between the producers and users of technology.

Significant features of the old S&T system (dating from the 1950s) that needed reform were

– The large number of institutes and S&T personnel at all levels of government in China;

– The lack of interaction between the institutes (because of their diverse reporting responsibilities) and the consequent duplication of activity (in Mission terms, "chaotic plurality"); and

– Their separation from the potential users of their outputs.

To illustrate the dimensions of this problem, in 1985 there were more than 4 900 independent R&D institutes affiliated with government above the county level (the county is an administrative unit of roughly 0.5 million people), with about 3 000 more at the county level. In the former group, more than 2 700 were engaged in industrial technology. It appears that the majority of the work done at these institutes was not R&D but was much more oriented to supporting SOEs that were weak in development capability and had little incentive to innovate. Paradoxically, this assisted such institutes in the reform process because, once given more freedom of operation, they were able to transform themselves into profitable businesses. The Mission saw several successful examples of this type of transformation at the provincial level (examples are discussed later in this chapter).

A large group of research institutes associated with CAS and a number associated with the central ministries and some provincial governments were engaged in strategic research. These were seen as the spearhead of national efforts to transfer high technology to the marketplace, with CAS leading the way.

Against the background, two significant trends have been

- The massive importation of technology from overseas, to rapidly increase the production capacity, as a direct consequence of the open-door policy; and

- The steady decrease of total expenditure on R&D by the government, leading to a national GERD of about 1.0% of GDP in 1988, 0.71% in 1992, and 0.67% in 1993.

Under the March 1985 Decision, this planned reduction in expenditure has been the instrument used to drive the reform process in the institutes. Government expenditures on R&D as a proportion of the state budget dropped by 14% between 1985 and 1986. Thus, from 1985 on, the government reduced its allocations to the institutes and forced them into the marketplace. This reduction was not uniform and depended on the type of institute and its affiliation:

- In the case of those institutes engaged in "public-good" R&D (such as health care, medicine, and population control), for which there is little room to obtain external funding, the reduction was only about 10–20% (of their baseline 1987 budgets);

- In the case of agriculture, which was seen as the highest priority for the country and in which again there is a strong element of public good, the reduction was only 10%;

- In the case of CAS institutes, where the government perceived a need to maintain a core of elite researchers engaged in fundamental research, the reduction was some 70%; and

- In the industrial-technology institutes, particularly in the provincial ones, the reduction was as much as 100%.

The timing of these reductions varied, with the more draconian measures being introduced fairly rapidly.

Combined with increased autonomy for institute directors and increased mobility for researchers, the effects have been far reaching, both in changing the culture of the institutes and in increasing their contribution to economic growth. This contribution is difficult to evaluate because of the influence of imported technology, but the Mission gained the impression that the contribution of high-technology industries to industrial output had increased from around 2% in 1985 to around 10% in 1995, with the expectation that it would be 15% by 2000.

These dramatic changes in funding have been complemented by other programs designed

- To maintain a capability in strategic research (for example, the 863 Program and the Climbing Program);
- To tighten the links to the application of R&D undertaken by the SOEs (for example, the National Key Laboratory Program and the concept of open laboratories); and
- To stimulate the setting up of new enterprises (for example, the Torch Program).

Concurrently, a technology market was created to legitimatize paid transactions in technology, to facilitate the links to application. This new concept was critical to the reform process because the former system provided no grounds for market transactions in technology.

Role of CAS in promoting high technology

As noted earlier, CAS was in the vanguard of the reform movement. With its 123 institutes and some 90 000 staff (of whom 50 000 were scientists and technicians, mainly in the natural and technical sciences), it constituted a significant force in the Chinese S&T system. Before 1984, CAS had virtually no interaction with industry or with the universities, which had a relatively low research base at that time.

However, a few far-sighted individuals had already started to challenge the old system. Thus, in 1980, the CAS Institute of Physics created the first technological entity that was not initiated, financed, or owned by the state — the Beijing Huaxia Guigu Information System Corporation Ltd (now a Chinese–US joint venture). Other enterprises of this type followed in Beijing; most were very small, but several rapidly became independent businesses, particularly in microelectronics. In Beijing, the Mission visited the San Huan Company (see Box 4), which had been developed in 1985 from the CAS Institute of Physics by a group working on rare-earth magnets.

These NTEs tended to cluster informally, initially in Beijing. In a parallel development, CAS in 1985 moved to formalize such clustering by creating the Shenzhen Science and Industry Park as a cooperative venture with the Shenzhen Municipal Government. The aim was to establish a base where high technology from CAS and other institutes could be combined with foreign investment and technology to open the way for commercializing high-technology products. This concept has given rise, over time, to the creation of 52 high-technology development zones throughout China.

CAS and its institutes have now created some 900 NTEs, of which about 500 are high-technology oriented and 400 are service oriented. Most are small, but a number clearly have potential to expand. The Mission was told that roughly 10% are very profitable; about 30% could be considered successful; another 30% are fair; and the rest are failures. This accords with the general experience in the countries of the Mission members. These CAS NTEs employ some 20 000 people: about 10 000 from CAS Institutes and another 10 000 on contract. This is a significant movement of staff, and the change in research culture can clearly be counted as a successful outcome of the reform process.

Concerns expressed to the Mission were that the CAS staff moving out into the NTEs were the brightest and most entrepreneurial scientists and engineers and that, although short-term gains may be made, there are potential long-term problems for Chinese basic research. These concerns have been acknowledged by SSTC with its programs to strengthen basic research. (In China, as in many other countries, a vigorous debate is under way concerning the appropriate role to be attributed to the universities in basic research.)

Again, CAS was farsighted in initially using some of its funding to promote basic research at the universities through a set of experimental measures that led to the creation of NNSF in 1986. By its system of competitive peer-reviewed grants, NNSF has stimulated research in the universities and also forced CAS scientists who seek support from it to be judged in the open market. Currently, CAS scientists get about 30% of the grants.

The Mission noted that CAS institutes in Shanghai and Guangzhou had decided to redirect their efforts to local problems and to the support of local priority areas. We were told that in Xi'an, the CAS institutes were not oriented to local problems. In Guangzhou, the Guangdong Academy of Sciences incorporates the 10 local CAS institutes, organized as the CAS Guangdong Branch, together with other local institutes. Such cooperation is a model for other areas, where the existence of unnecessarily large numbers of separate institutions has led to duplication and inefficient use of research resources. In Shanghai, the Shanghai Academy of Sciences, including institutes of central and local government departments, has been set up to give a coherent approach; there is a clear opportunity for the local CAS institutes of the CAS Shanghai Branch to participate, and from the Mission's perspective this would seem to be desirable.

It was suggested to the Mission that the CAS institute structure be dissolved and the institutes handed over entirely to local governments to concentrate on local issues. The experiences of the Mission members suggest that this would not be in the national interest in the long term — although major restructuring may be necessary to streamline and redefine the mission of CAS, a core of CAS institutes provides the national strategic-research resource that is essential for China's future development. The formation of national centres, such as the Centre for Life Sciences in Shanghai, funded by CAS, SSTC, and universities, exemplifies a new approach to the task of making more efficient use of research resources.

The Torch Program and high-technology development zones

The Torch Program was launched in 1988 by SSTC to build on the concept of technology parks initiated by CAS. Given the sluggish response by SOEs in accepting new technology (either through transfer of technology or by incorporating R&D institutes into existing enterprises), the technology parks were seen as a new and dynamic way to bring high technology from the research institutes into the marketplace.

The Torch Program had as its major thrust the establishment of a number of high-technology development zones for NTEs. This was supported by preferential tax treatment in the zone (15%, instead of the national rate of 33%), special loans to finance new enterprises (various schemes were developed by provincial governments to encourage banks to cooperate), and preferential treatment for designated high-technology industries (essentially, those defined in the 1978 National Conference on Science and Technology, plus information technology, mechatronics, and ocean engineering).

The number of zones has increased rapidly since 1988, and there are now 52 national-level zones with a wide geographical spread. These contain 10 000 NTEs employing 80 000 personnel and producing 80 million USD worth of goods. This growth has been accompanied by the development of 60 incubator zones to assist individual entrepreneurs seeking to establish new enterprises. Not surprisingly, there are considerable differences between the zones, depending on their location and on the degree of support given by the local government. Some are located in urban areas with concentrations of R&D institutes and universities and are based on existing infrastructure, with the aim of exploiting the potential of the available S&T expertise. The Mission learned of such zones in Beijing and Shengyang (Table 11).

Others include zones located in existing or planned industrial zones with clearly defined boundaries and extensive new infrastructure, where there is encouragement for scaling up high-technology enterprises based on imported technology and capital. Five special economic zones and the Pudong Development Area include sites allocated to technology-intensive projects. In the case of the Pudong Development Area, there is also a designated high-technology development zone (Zhanjiang High-technology

Table 11. Cities with zones for NTEs visited by the Mission.

Zone	NTEs (n)		Turnover (billion CNY)		Employment (n)	
Beijing	4 000	(1995)	7	(1995)	29 000	(1991)
			25	(2000)		
Shanghai	28	(1991)	0.7	(1991)	18 700	(1991)
Shengyang	250	(1991)	0.36	(1991)	5 800	(1991)
	600	(1991)				
Anshan	>150	(1995)	0.7	(1995)		
Guangzhou	185	(1991)	1.44	(1991)	2 600	(1991)
Xi'an	200	(1995)	2.5	(1995)		

Note: In 1996, 8.07 yuan renminbi (CNY) = 1 United States dollar. NTE, new technology enterprise.

Park) in the same area. The Mission had the opportunity to visit several such zones at Pudong, Xi'an, and Anshan and was impressed by the scale of the developments and the range of industries located there (see Table 11).

The fact that it is becoming difficult to separate the exact roles of the activities in the development zones testifies to the success of the reform to set up the high-technology industry as a significant industrial sector in China. The large concentration of imported technology is clearly a significant factor in that success, but there is also a notable contribution from domestic NTEs that have spun off from research institutes. However, there is a need to develop more linkages between the NTEs and overseas companies located in the zones, possibly through new joint-venture operations. It is likely that the preferential policies enjoyed by such zones will have to be phased out when China joins WTO, but the Mission believes that the dynamism of the zones, as evident during their visits, will ensure their growth.

NTEs and TVEs

During its visits the Mission saw a number of different types of NTEs that had spun off from research institutes. Their characteristics depended on the parent institution, the local government, and the nature of the technology involved in the spin-off process. The Mission had an opportunity to see only some examples of a much more diverse group, but these had interesting lessons for the future. This is clearly a rich field for study by policy analysts, and the Mission would encourage more detailed studies.

Form 1: Part of an institute spins off as an independent NTE

The first form of NTE seems to be associated mainly with spin-offs from CAS institutes and universities. It involves a piece of the organized structure (human-resource, technological, and, in many cases, physical assets) of the original institution being diverted to establish a new business entity. This ensures that the NTE has a relatively strong starting point, although there is no guarantee that this will ensure a successful commercial operation.

The Mission was told of various examples of successful spin-offs and had the opportunity in Beijing to visit the San Huan Company (see case study, Box 4), which is an excellent example of a CAS institute spin-off initially funded by internal funds from CAS. Other CAS groups have spun off to form joint ventures — with CAS and provincial governments both providing funding (for example, integrated-circuit production in Shanghai) or with foreign companies providing capital and equipment to complement CAS inputs of people and technology (such as magnetic-resonance imaging systems in Wushi).

Box 4

Form 1: San Huan Company, Beijing

San Huan Company began in 1985, when a group of professionals working on rare-earth magnets spun off from the CAS Institute of Physics. Initially, there were problems with patent rights involving Sumitomo in Japan and General Motors in the United States, and production was restricted to the domestic market. In 1993, these problems were resolved, and the company opened up export markets through a subsidiary, San Huan International Trading Company. Production rose from 10 t/year in 1985 to 230 t/year in 1994 (with a turnover of 120 million CNY). The ownership is 84% CAS and 16% Institute of Physics.

San Huan has set up six subsidiary companies to increase production and to develop applications for rare-earth magnets in electric vehicles and scientific instruments. It has a worldwide distribution network and is the fourth largest rare-earth-magnet company in the world. Currently, there are 800 workers employed in the group of San Huan companies, including 100 at the Beijing headquarters. Of the latter, roughly half have come from the CAS Institutes of Physics, Electrical Engineering, and Electronics. The company has its own research group of 10 people working on improved and new rare-earth magnetic materials.

Form 2: An institute transforms itself into an independent NTE

The second form of NTE seems to be associated mainly with provincial industrial-technology institutions that have had their funds completely cut off over a period of 3 years and in consequence have been forced to change their mode of operation completely. This transition has been particularly difficult because the SOEs that formerly drew on the institutions' services for free have been reluctant to pay market price for their technologies. Such transfers of technology are often complicated by disputes over IPR, which institutes are now seeking to enforce. In this regard, the technology market has not yet begun to operate effectively.

The Mission was given several excellent briefings by chief executives of such organizations in Shengyang, Anshan, and Shanghai. The examples covered a range of industrial sectors:

- In Shengyang, electronics, machinery (see case study, Box 7), and building materials;

- In Anshan, electrostatic technology; and

- In Shanghai, light industry (see case study, Box 5) and organic new materials.

These discussions brought out a number of common themes:

- The success of the transition depends to a very large extent on the entrepreneurial ability of the chief executive;

- The dual leadership role imposed on the chief executive, with the heavy burden of social responsibility for his or her workers (both current and retired) coupled with technical leadership, causes many tensions;

- The failure to market technology to local SOEs forces the transformed enterprise to seek markets elsewhere in China and overseas; and

- National recognition in some form is a considerable asset in the marketplace.

Form 3: S&T personnel move as individuals to create independent NTEs

The third form of NTE arises when scientists and engineers move, individually, from their previous R&D institutes. They move out with embodied knowledge as their primary resource, together with experience in R&D and design and innovative ideas. Their basic problems arise in obtaining finance, although this is becoming easier as provincial authorities set up

Box 5

Form 2: Shanghai Light Industry Research
Institute, Shanghai

Established in 1958 as a research laboratory by the Shanghai Municipal Government, the Shanghai Light Industry Research Institute originally carried out research on behalf of light industries that manufactured cosmetics, food, watches, cameras, textiles, glassware, and so on.

This institute, just like all others in China, is state owned. Before the S&T-system reform, its budget allocation consisted of core funding for salaries and benefits and project funding for researchers and was provided entirely by the government. It was staffed with bench-top researchers doing laboratory work without even a pilot plant for research results to be developed into prototypes. During the reform process, the core funding was reduced to zero by 1991 (except for pensions for those already retired), and the project funding was totally cut off by 1993. However, despite the adverse conditions faced by the Institute during the initial phase of the transformation, it has blossomed in the past 10 years.

The value of the annual business dealt with by the Institute rose from 1.17 million CNY in 1985 to 19.89 million CNY in 1994; net profit rose from 1 million CNY in 1985 to about 5 million CNY in 1994. The number of staff has been reduced from 309 in 1985 to the current level of 210, but productivity increased from 3 800 CNY per employee in 1985 to 23 000 CNY per employee in 1995. In 1991, the Institute began to set up its first pilot plant for the development of chemical products, an investment of about 5 million CNY, and in 1993, it invested in another pilot plant for mechanoelectronic processes. Furthermore, in 1992 the Institute terminated its traditional role of providing subsidized housing to new employees, who are now assisted in purchasing their own accommodations. All staff are on contracts of various durations. Before the reform, the Institute widely practiced reverse engineering, but now it pays much attention to IPR and severely penalizes its staff for infringements, even prosecuting them in court. It also trains its staff in ethical business practices.

According to the director of the Institute (who had taken courses in modern management techniques at McMaster University in Canada under funding from the Canadian International Development Agency), the key to the successful transformation of the Institute during the reform in the past 10 years was innovation. The entire Institute was engaged first of all in attitudinal innovation, which in turn led to technical innovations, organizational innovations, and innovations in personnel management.

mechanisms to give start-up grants and to assist with low-interest loans. In Guangdong, the Mission was told of the activities of the Guangzhou Technology Corp. Ltd, set up by the Guangzhou STC.

Unfortunately, the Mission did not have an opportunity to visit any such companies, but we were told in Shengyang that some 60% of the 600 or so NTEs in the Nangpo high-technology development zone had been set up by individual researchers, largely from the local CAS institutes and universities. The role of the universities was reinforced during discussions in Shanghai, where Shanghai Jiaotong University indicated that there were 168 spin-off companies founded by its staff, with sizes ranging from 4 or 5 to 300 people. Here the university has assisted with venture capital and management and claims to be annually receiving 100 million CNY in revenue from its investments. Three of the more successful of these spin-offs have gone public to seek further capital for expansion.

Form 4: An institute forms a linkage with a TVE

Research institutes have developed an interesting range of linkages with TVEs. Given that TVEs overall produce between 40 and 50% of the total economic output of China and that in many cases they have moved rapidly into many advanced-technology areas — often well ahead of SOEs — these linkages deserve special consideration. For the new TVEs, the pressure to compete is driving them toward alliances with research institutes. For their own survival and development, the research institutes are driven toward the TVEs as willing and receptive partners in innovation.

The range of products and processes covered by TVEs is not limited to the production of agricultural products. As the Mission learned in its discussions, depending on location, 50–70% of the TVEs are engaged in textile production, building materials, plastics, and electronic components. It appears that TVEs are often small but dynamic (because of their collective ownership), and so they are better able than the larger and more conservative SOEs to sense market demands and opportunities to use technology developed by the institutes. The links are often established by individual researchers from institutes seeking to augment their salaries through external contracts or family linkages. The tacit transfer of knowledge can in this way lead to issues of ownership of intellectual property, but it can also lead to the growth of TVEs. Xiangyang Chemical Factory in Liaoning Province is one of several successful examples described to the Mission (see case study, Box 6).

> Box 6
>
> **Form 4: Xiangyang Chemical Factory, Liaoning Province**
>
> The Xiangyang Chemical Factory was set up in 1982, with seven employees and capital of less than 1 000 CNY, as a TVE in Yingkou Township to manufacture wall paint. In 1987, it made contact with the CAS Institute of Chemistry in Beijing as a result of the links of a professor there with the village. The factory and the Institute jointly developed powder coating materials for domestic appliances, and with the resulting high-quality product, production and profits grew rapidly (from 650 000 CNY in 1987 to 14 million CNY in 1989).
>
> In 1989, the factory and the Institute jointly established a new company to manufacture a new organic catalyst developed by the Institute. Production grew rapidly to 8 t of catalyst in 1995, with a sales value of 30 million CNY. The factory now employs 262 people and has fixed-asset capital of 10 million CNY. It is planning to expand production to 20 t per year, using investment capital from a bank loan of 6 million CNY and self-funding of 4 million CNY. The Institute is developing new catalysts for a contract fee of 600 000 CNY per year.

The 863 Program

Although the Mission did not have an opportunity to see any practical aspects of the 863 Program (so called because it was established in March 1986 as a result of a proposal by four eminent scholars), it is important to briefly describe it to illustrate the degree of China's commitment to high technology. The 863 Program is a medium- to long-term plan for the development of high technology in China. It stemmed from the concerns of researchers to maintain a high-quality fundamental research program at a time when policy was heavily oriented toward the application of S&T for economic growth. The resulting activities would be referred to as strategic research in industrialized countries.

The 863 Program emphasizes development of start-of-the-art technologies in seven areas: biotechnology, space technology, information technology, laser technology, automation, new energy sources, and new materials (these are similar to the areas identified in the 1978 National Conference on Science and Technology). With guidance and funding from

SSTC, more than 13 000 scientists and engineers are engaged in the program through a series of networks and a number of newly established research centres. These include the Computer Integrated Manufacturing Systems ERC, the Intelligent Robot Research Centre, the Photo-electron Research Centre, the Genetic Engineering Drug Research Centre, the Artificial Crystal Research Centre, the Genetic Engineering Biological Products Research Centre, and the Genetic Engineering Vaccine Research Centre. The relationship of these centres to the many other specialized research centres in the country is not clear to the Mission, but there appears to be substantial overlap in a number of fields.

Five of the seven areas are under an area office located in SSTC (the other two fall under the jurisdiction of the Commission for Science, Technology and Industry for National Defence), and activities are coordinated through an expert committee for each area drawn from leading scientists in CAS, the universities, and appropriate government institutes. Emphasis in the 863 Program is on

- Bringing the brightest young researchers in China together in the centres;
- Encouraging Chinese scientists overseas to return to the centres; and
- Using the centres as active links to international S&T.

Chapter 8

STATE-OWNED ENTERPRISES

The SOE sector

The SOE sector in China constitutes an enormous share of the national economy, accounting for about 43% of China's gross value of industrial output (down from 78% at the beginning of the reform period in 1978).[9] The SOEs still provide some 70% of the industrial jobs at or above the county level in China, despite the rapid growth in employment opportunities associated with the TVEs.

Of more than 9 million Chinese industrial enterprises (the great majority of which are private), some 104 700 are SOEs, most of which are controlled by provincial- or subprovincial-level governments. About 14 000 of the SOEs are classified as large and medium-sized enterprises. These form the core of the state industrial sector, accounting for about 80% of the total SOE output and 35% of the total industrial output. Of these, 2 000-3 000 are considered very large. SOEs supply the major share of tax revenues to different levels of government; in 1993, this accounted for 65% of government revenues.

Although not all SOEs are unprofitable (see Box 7 for a description of a successful one), it is likely that as many as two-thirds are. In 1993, losses from the SOE sector were estimated at 2.4% of GDP for SOEs of all kinds

---

[9] Unless otherwise noted, this background discussion of SOEs is drawn from Broadman (1995).

> Box 7
>
> **Shenyan Blower Works — a successful SOE**
>
> Shenyan Blower Works is one of the world's top 10 producers of turbine compressors, blowers, and fans. It has recently attained ISO-9000 certification. In association with a CAS institute and six Chinese universities, the enterprise has developed and put into operation a highly successful computer-integrated manufacturing system (CIMS). Shenyan Blower Works does its own product design and develops highly sophisticated software for its CIMS with its Chinese partners, but it imports the most important computerized machine tools from industrialized countries.
>
> Shenyan Blower Works pays great attention to worker training, sending staff members abroad each year to become accustomed to the most modern techniques; the entry level for employment is equal to college-entry level, and all of its machine operators get a 3-year training course, plus regular updates. This SOE demands high-quality products from its suppliers and even purchased castings from the United States when locally fabricated castings failed to meet its specifications. A key element of this enterprise's success has been its entrepreneurial management.

(down from 5.3% in 1990). For SOEs in the industrial sector, the figure for 1993 was 1.4% of GDP, down from 2% in 1990. The state's explicit subsidies for these losses have been declining. However, there has been a growth of implicit, "quasi-fiscal" subsidies in recent years — typically, "policy loans" through the People's Bank of China — and these are thought to have grown to 5.1–6.8% of GDP by 1993.

In the period since the initiation of reforms, many large and medium-sized SOEs have improved their management and technology as a result of technology imports, greater attention to domestic sources of technology (especially in-house R&D), and the pressure of market competition. Those that have taken the business of reform most seriously seem to be the ones whose technology levels have improved, although other factors (such as firm size, industry type, and location) also seem to have been important. However, according to one recent study, industrial growth still seems to be driven mainly by new inputs of capital and labour; the old pattern of extensive growth still seems to prevail. This pattern of growth is no longer sustainable and will become even less so once China becomes a member of WTO (STC–CPPCC 1994).

Until quite recently, SOEs continued to operate according to a planning system and were confident that the state would cover their losses. They were still giving bonuses to their workers and continuing to neglect innovation, even in the face of losses. Policies for industrial technology have not been carried out. For instance, SOEs are expected to devote 1% of sales to technological improvements; few do. The lack of commitment to technological improvement has also induced capable technical people to leave the SOE sector for NTEs, TVEs, joint ventures, and wholly foreign-owned companies. The movement in personnel can also be understood in terms of the reward structure for scientists and engineers: status and pay are low, and according to one survey, bonuses are typically 15% lower than those for ordinary workers (STC-CPPCC 1994).

An important issue appears to be the types of technology that should receive attention. Much focus has been given to advanced technology, and this has led to parts of the industrial-technology system of an enterprise becoming fairly modern. However, attendant technologies (such as for parts and components) often lag behind. Another issue is whether an industry should strive for a general technology upgrading, with efforts to bring other enterprises up to the level of industry leaders (through pro-diffusion policies, for instance), or whether the strengths of the leaders should be enhanced instead, to meet international competition.

SOEs are products of the pre-reform planning system and the regulatory and management policies of that earlier era. At the beginning of the reforms, most SOEs had suffered from policies and management styles that discouraged technological innovation. As a result, most were characterized by seriously outdated technologies. In addition, most SOEs carry heavy social-service burdens and have redundant labour forces, typically 30–40% greater than needed. Recent studies have estimated that social services constitute about 6% of the total costs of the average urban SOE and account for 40% of its wage bill. These heavy social-service burdens make it difficult to achieve enterprise reform and also help explain why the state is reluctant to allow SOEs to wither and die.

At the Anshan Iron and Steel Works (Angang), for instance, total staff now numbers about 190 000 (Table 12). Angang's restructuring plans call for the shedding of ancillary enterprises and social-service organizations, which will reduce staff in the core steel-making business to about 70 000. If its annual steel-making capacity is raised to the projected 9 million t, it will still be competing against international firms that on average produce the same amount of steel with a staff of about 15 000. It is clear from the Angang case — admittedly an unusually large and complex SOE — that many of the problems with SOEs are not, in the first instance, technological ones. Indeed, before modern technologies can "hope" to improve the performance of many SOEs, the larger problems of managerial, structural,

Table 12. Structure and work force of the Anshan Iron and Steel Complex.

Company element	Work force
Iron and steel production	70 000
2 iron-ore mining companies	30 000
Mechanical and machine building	15 000
Electrical manufacturing	10 000
2 construction companies	20 000
Railway transport	10 000
7 research institutes	10 000
Staff welfare organizations	N/A
8 hospitals	10 000
Education (up to college)	9 000
Housing system	N/A
Policing for Anshan	N/A
Total complex	190 000

Note: N/A, not available.

and social reforms must first be solved. As one recent analyst (Zu Zhaoxiang 1995, p. 83) put it,

> Preliminary experience drawn from reform practices has shown that a lack of technological dynamism is a manifestation of a much larger structural problem that cannot be solved by any shortcut or through any simple policy instrument alone. Successful industrial technology development is a result of the interplay of many factors.

For many SOEs (perhaps one-third), the chances of survival are so marginal that the state is prepared to allow them to go bankrupt or to be sold. For perhaps another one-third that are also losing money, there seems to be a commitment to save them. For the one-third or so profit-making enterprises, the challenge is to upgrade them through managerial reform and technological change and make them internationally competitive.

Necessarily, then, SOEs loom large on the state's agenda. Given the variety of firm sizes and the variety of industries they represent and given the great social and economic complexity that changing them represents, identifying and making the right policy choices to deal with SOE problems have not been easy. Although the strategies continue to evolve, they have involved the following elements:

— Ongoing waves of enterprise-management reforms,

— Broader economic and social reforms pertaining to the environment in which the enterprises operate;

— Ambitious efforts to upgrade the technology of enterprises through the procurement (often from abroad) of advanced production technology; and

— S&T reforms intended to bring Chinese R&D efforts to bear on the SOEs' technical needs.

These strategies are interrelated, so it is difficult to discuss S&T reforms pertaining to SOEs without also discussing the other elements in the strategy.

This is not the place to attempt a complete review of the reform environment of SOEs. Suffice it to say that there has been an evolution in reform thinking, from early efforts to enhance managerial authority while gradually introducing market mechanisms, to the introduction of the contract-responsibility system, to the growing realization of the importance of economic legislation, to a realization of the importance of clarifying property rights, which in turn required a more careful inventory of state assets. With better accounting of assets, attention began to turn to control and mobility of these assets and to a shareholding system. The 1992 *Regulations on Transforming the Management Mechanisms of State-owned Industrial Enterprises* (State Council 1995) nominally extended 14 new rights to enterprises to enhance their autonomy and authority. Relatively few enterprises have in actuality been ceded all 14 rights, but the downward devolution of authority has nevertheless been substantial, as was evident from discussions with the Mission.

With the March 1993 introduction of the concept of socialist market economy as the guiding principle of economic philosophy and with the Decision on Issues Concerning the Establishment of a Socialist Market Economic Structure, made at the Third Plenary Session of the 14th Party Congress in November 1993, a new surge of reforms is transforming the context in which discussions about the future of SOEs should occur. In particular, policy support for the idea of a modern enterprise system is opening up possibilities for diverse ownership schemes; creating significant industrial reorganization, leading to large national corporations and conglomerates; and changing relations between government and industry, which are intended to separate state-ownership rights from the operational property rights of the enterprise or corporation as a legal person. For such reforms to go forward requires additional reforms in macroeconomic and monetary policy, further expansion and liberalization of labour and capital markets, the implementation of a social-security system, and the full implementation of a legal system suitable to a market economy.

In 1994, 100 enterprises were selected for full corporatization within 2 years. Under the leadership of SETC and SCRES, this corporatization will involve the conversion of selected enterprises into limited-liability companies and limited-liability shareholding companies, settlement of their outstanding debts, concessional bank financing, income-tax concessions, and a reduction of their social-service burdens (Broadman 1995). At least one of the enterprises visited by the Mission (the Yanshan Petrochemical Plant) is included in this experiment.

The changes noted above were going on during the Mission but were not its primary focus. What was clear was that there were fundamental changes at work in some of the enterprises visited and that these were transforming the challenges of technological innovation and R&D in fundamental ways. Enterprise reform was not the primary concern of the Mission; as a result, we did not have meetings with the appropriate agencies charged with the cardinal restructuring tasks.

S&T reforms and the SOE

What was clear, however, was that given the enormous implications that technological upgrading and industrial restructuring have for national S&T policy (nominally, the domain of SSTC) and given that these tasks fall more into the hands of the other commissions (especially SETC and SCRES), it was understandable that a number of our interlocutors bemoaned the fact that national S&T policy is fragmented. It was also clear that as Chinese enterprises are being restructured and reorganized and their relations with the state change, S&T is coming to be correlated with the needs of Chinese industry in ways that are more reminiscent of firms in a market economy than of enterprises in a socialist-planned economy. This fact again points to the quandary of how to approach the development of policy and coordinating mechanisms within the national government for a "marketizing" economy in which R&D spending by enterprises is expected to become the main source of GERD.

In addition to domestic economic and management reforms as factors shaping S&T in SOEs, the open-door policy has had a major impact on the SOE sector. This impact comes in two forms:

- With Chinese enterprises increasingly participating in the global economy and with the growth of foreign business activity in the Chinese domestic market as a result of the gradual liberalization of Chinese foreign-trade policies, SOEs are facing market competition in ways that never occurred in the pre-reform period. This has begun to create incentives for improving enterprise technology and, thus, to reverse the attitudes and behaviour of the pre-reform period. The pace of this change is often slower than might be desirable, but it was evident in all the firms visited by the Mission.

- In conjunction with the first point and with other reform policies, the open-door policy has led to the acquisition by Chinese enterprises of massive amounts of technology from abroad. Shanghai alone has spent about 20 billion USD on imported technology since the late 1970s; the estimate for the whole country is more than 40 billion USD. This surge of imported technology, plus the technology upgrades from domestic sources, has made the technological level in many SOEs considerably more appropriate for market competition than had been the case before these programs of "technological renovation" (*jishu gaizao*) were begun. Still, on average, the technological level of SOEs is well behind 1990s international standards, and the need for ongoing technological upgrading continues to be a pressing matter.

Examples of successful technology enhancement in China viewed by the mission (and that are consistent with experience in other countries) have involved the introduction of technology from abroad, which was then complemented by local R&D to facilitate the full and effective assimilation of the imported technology and to provide for the complementary technologies and technical services that the major piece of technology from abroad would require. Although successful transfers in China have had profound effects on the quality, price, and overall competitiveness of Chinese industrial products, there is also evidence that there has been less attention to this area of policy than there might have been. It has been observed, for instance, that, on average, for every 100 CNY spent on procuring technology from abroad during the 1980s, only 9 CNY was spent on absorption and assimilation. This is in marked contrast to South Korea and Japan, where major R&D efforts were made to assimilate imported technology (STC–CPPCC 1994).

The pre-reform technological backwardness in SOEs was not only a function of an economic system that discouraged innovation, as noted above, but also a function of the design of the S&T system. To this day, most SOEs are organized and led by production ministries (or production bureaus at local levels). These ministries also have their own free-standing R&D institutes, supposedly to serve the enterprises under their ministry, as in the Soviet model. Most of the 5 000 or so Chinese research institutes at the county level or above — and thus the workplaces of most of China's scientists and engineers in R&D — are institutes of this sort. Some very large enterprises have had their own in-house R&D capabilities, and as a result of the reform policies many other large and medium-sized enterprises are beginning to establish them, but most SOEs do not. The intention was that they would rely on the work of the free-standing ministerial (and industrial-bureau) institutes to meet their S&T needs. Similarly, technical personnel, for both research and production, were to come from specialized institutions of higher education, which were also part of the ministerial system (*xitong*).

A chronic problem with this system was the failure of the research subsystem to be linked to the production subsystem. As a result, technology in enterprises was not being infused with new ideas from the laboratory and new technologies from pilot plants; research at the institutes stagnated, having neither effective demand from the economy for new research directions nor a stream of new personnel from outside the closed ministerial system. Generally, two very different cultures grew up in the two subsystems, and this worked against collaboration and cooperation.

As seen in Chapter 4, a central objective of the S&T reforms has been to introduce a number of reform policies to overcome this separation of research from production. These efforts included the introduction of a labour market for scientists and engineers and of technology markets and contract research (requiring in turn a fundamental reconceptualization of intellectual property and the establishment of a patent system). More important, however, the introduction of market competition via the more general economic reforms began to stimulate demand for new technology, and the major changes in the funding of research (and the diminution of guaranteed appropriations to research institutes) have forced a fundamental change in the character and behaviour of research institutes. Despite some successes, the widespread persistence of many of these problems in the mid-1990s indicates that reform efforts have not been fully successful. Given the complexity of the problems, this should not be surprising.

Although the technology markets and contract research indicate that some progress has been made in overcoming the separation of research and production, it appeared to the Mission that far more fundamental changes were going on that were made possible by the reform environment but were not entirely anticipated. The S&T reforms, especially funding reforms, seem to have had a greater impact on the institutes than on SOEs. The initial experience with technology markets in the late 1980s seemed to be troubled with the inherent difficulties of placing a value on technology and with the weakness of demand for technology coming from enterprises that until quite recently had yet to face up to the realities of market competition. In the face of such problems, China used another policy to encourage the merger of enterprises and research institutes. The diverging cultures of the two types of organizations and the difficulties of managing all the social-welfare implications of such mergers have limited the effectiveness of this policy as well.

However, the acceleration of economic and management reforms and the widening of the open door in the early 1990s have created a new environment for linking research to production and, most important, have created conditions for a variety of institutional innovations that promise considerable benefits for Chinese industry as a whole, including SOEs. On the basis of the Mission's observations, these innovations fall into three categories:

- The enterprization (*qiyehua*) of research institutes — During the course of the Mission, we had opportunities to observe the ways a number of industrial institutes (under either central ministries or local industrial bureaus) had responded to the drastic budget cuts they had experienced and to the new market environment (see Box 8). Although the cases we saw were of successful institutional innovation, they clearly indicate that under the right reform conditions, reform objectives can be met. Not all of these cases involved the formal transformation from a research institute to a company (*gongsihua*), but in all cases, the organization involved had found an economically useful role — as consultancy, technology broker, engineering services provider, manufacturer, etc. — that did not exist but often was needed in the pre-reform, SOE-dominated industrial

Box 8

Liaoning Provincial Machinery Research Institute — from institute to "virtual corporation"

In 1990, the Liaoning Provincial Machinery Research Institute's fixed assets were evaluated at 6.7 million CNY, and the government's contribution to its budget was reduced to zero. Today, the Institute's fixed assets are estimated at 31 million CNY, and it pays salaries far above the provincial and national norm (3 000 USD per year on average, with the top scientist earning 20 000 USD).

In 1995, the Institute derived 50% of its income from the sale of equipment it designed for a variety of industries (such as equipment for food processing, environmental protection, and air conditioning) and expected this share to go up to 60% the next year. The Institute's other income is from R&D contracts and technology licencing. It has had no success competing with national institutes for national contracts and so deals almost exclusively with enterprises.

The Institute manufactures in its own pilot plant and factory (which employs 200 people) and through an extensive system of subcontracting to local factories, some of which are completely dependent on the Institute for their continued existence. (The Institute would never consider buying an SOE because of the burden of social-welfare costs it would acquire.) The Institute is now establishing its own small offices overseas (for example, in Kuala Lumpur) to explore foreign-market opportunities.

The local Liaoning Machinery Bureau still has the right to appoint the Institute's director and can require the Institute to pass funds to the Bureau.

economy. Common elements of success seem to include technical capability, a supportive policy and political environment created by local authorities, and, most important, entrepreneurial leadership.

- Joint ventures between research institutes and factories — Joint ventures are based on clear market objectives and market needs and on a realistic understanding of the value of the assets that each party contributes to the new venture. Enhanced market intelligence and progress in thinking about intellectual property and shareholding systems make ventures like these profitable and sustainable in the 1990s in ways that were not possible in the 1980s.

- Open approaches to technology acquisition — Cases in which technology seems to be making a difference in the economic vitality of SOEs are characterized by very open and diverse approaches to the acquisition of technology. Except for firms that have very specialized technologies developed in China, most of the successes rely heavily on imported technology but supplement this with Chinese technology and their own R&D. Special efforts to assimilate the technology pay off in launching the enterprise on a more sustainable trajectory of positive technological change and creating an enterprise culture that welcomes and appreciates technological innovation. These firms typically already spend more than the national average on R&D. Although in-house research is going on, success also seems to be characterized by participation in networks of R&D activities that bring together complementary talents from the firm itself and from research institutes and institutions of higher education from both inside and outside the ministerial system. These open approaches — mixing make-buy, search-research, and domestic-foreign distinctions in new ways — are characteristic of international trends; the participation in research networks of different organizations with complementary S&T capabilities is consistent with current thinking about new approaches to the production of knowledge.

These impressions of apparently successful cases have implications for future policy development. At stake here is the implementation of a modern national system of technological innovation to replace one built on faulty or outdated assumptions. The implementation of such a system, however, depends on a number of factors beyond the normal reach of S&T policies. These include the broader context of economic reforms (especially with regard to prices, taxation, IPR and other areas of business and commercial law, foreign trade, etc.), industrial structure, and enterprise organization and management. In light of these factors and the size and diversity of the overall industrial economy, it is very difficult to design a single set of S&T policies — including reform policies — for SOEs as a whole.

It is clear that many of the S&T reforms and policy initiatives are going in the right direction. In addition to those noted above, the recent SSTC programs for industrial extension, technology diffusion, and nationally linked productivity centres are likely to produce important benefits, especially to small and medium-sized firms. But as with other programs, they are premised on enterprises' giving rise to effective demands for new technology.

Issues for the future

If China's GERD–GDP ratio is to reach the 1.5% figure that has been targeted, SOE spending on R&D will have to increase significantly. This raises a number of issues:

— It is likely that government policy will be required to induce enterprises to meet the increased expectations placed on them. The Mission heard mention of special tax incentives, for instance. It is clear, however, that there is a danger that such incentives will mean little more than increased tax expenditures by government, rather than increased R&D expenditures by enterprises. The trick in designing incentive programs is to induce incremental spending by the enterprise sector.

— If the enterprises can be encouraged to increase their R&D expenditures, who will do the work? Increased in-house research would, in principle, be desirable. However, as relatively few SOEs now have their own R&D facilities and traditions, it may be difficult for them to organize and staff up to do high-quality work during the period of the Ninth Five-Year Plan. This suggests a need for highly innovative approaches for SOEs, such as working in networks of researchers and innovators outside the enterprise. We have already identified this pattern as one that is working in China in some successful cases. It may imply a need to give special attention to the management of innovation in enterprises, including, but not limited to, high-quality in-house research.

— The development of R&D strategies for SOEs will be occurring in a highly fluid and unsettled organizational environment brought on by corporatization. There is likely to be a great deal of confusion about ownership of all sorts of assets, including intellectual property. Should intellectual property be considered as belonging to a company, to a conglomerate, to a supervising ministry, or to a holding company? It is quite possible that ambiguity over IPR in the context of organizational change could significantly deter the expansion of SOE expenditures on R&D enterprises.

— The appeal of acquiring advanced foreign technologies, either through Sino–foreign joint ventures or independently, will continue to be strong. In some ways (though not all), these technology-acquisition strategies have worked to the advantage of SOEs but against the objectives of domestic R&D policies. This is likely to be the case in the future as well. To achieve a balance between the advantages of technology transfers from abroad and the advantages of domestic R&D, there is a need for sophisticated policy mechanisms for coordinating technology imports and R&D. In part, these are needed to facilitate the absorption and assimilation of foreign technology but, perhaps more important, to set realistic priorities to ensure that R&D resources are not wasted on projects that, even if successful, yield technologies that would be behind those already available at the international level. Such mechanisms will require new forms of government–business cooperation to ensure that both the short–term needs and perspectives of firms facing the market are reconciled with the concerns of the policymaker and the research community for the S&T well-being of the nation over the longer term.

— The NTEs and research institutes that have successfully suffered through the agonies of reform and are now profit-making enterprises represent new forces in China that can provide SOEs with the technological and engineering services needed to fill gaps between the productive and engineering capacities of SOEs and the needs of customers. These new creations can also be seen as possible competitors of SOEs in some industries and as having interests in S&T-policy outcomes that are different from the interests of SOEs. A policy environment that will maintain creative competitive tensions while also stimulating useful synergies between the two sectors will be a challenging but worthy policy objective.

As noted above, the challenges posed by the SOE sector go far beyond the normal concerns of S&T policy. Although the latter should focus on aiding the transition of SOEs to profitability through technology-driven productivity gains and new product mixes, there is also something to be said for a proactive stance for S&T policy in the SOE sector. From this perspective, the attention is less on the transitional issues than on what to do with S&T in the successfully reformed SOEs. Because it is assumed that these will be market-competitive firms with high degrees of managerial autonomy, fresh thinking about their S&T needs will be required. New terms of reference will be required, with concepts like corporate R&D and innovation management replacing the concept of SOE S&T activities, and the idiom of policy will need to shift toward the complexities of government–business relations in a market economy.

Chapter 9

AGRICULTURAL RESEARCH AND RURAL DEVELOPMENT

Agriculture, which for the purpose of this discussion includes forestry and fisheries, is China's most important economic sector — in 1990 it accounted for 20% of GDP and 60% of the total employment in the country. With only 7% of the world's arable land, China has been responsible for feeding 22% of the world's people. The performance of this sector is therefore vital to China's food security and is the basis of the living standard of a major proportion of the country's population. Over the last 30 years, despite the difficulties experienced by the Chinese agricultural research system, investments in research have been responsible for 20% of the total increase in agricultural productivity (Fan and Pardy 1992). For all these reasons the further development of S&T in agriculture, by strengthening research and improving the transfer of technology to the end-user, was given a high priority in the Eighth Five-Year Plan and in the significant policy events leading to the reform of China's S&T system.

Policy agenda for agriculture

After the May 1995 Decision on Accelerating Scientific and Technological Progress, the new S&T policy has as one of its major objectives the development of a more efficient, modern agricultural system, through the application of advanced equipment and agrotechniques, to meet China's future food requirements. This will be achieved by

- Making S&T progress a high priority in the development of agriculture and the rural economy;

- Integrating agricultural science with education (popularizing the achievements of agricultural S&T) and improving the integration of R&D with industry;

- Strengthening the quality and relevance of S&T in agriculture and improving the scientific training, equipment, and personnel policies of the research staff; and

- Promoting S&T in TVEs to accelerate the use of modern agricultural S&T management and helping them to develop new technology-intensive industries that also broaden the opportunities for employing surplus rural labour. (Such technologies should also be extended as rapidly as possible to help the farmers in rural areas where poverty is endemic.)

In addition to these efforts, the policy recommends that the complex organization of agricultural R&D in the country be restructured and that its management be improved to produce a more output-orientated research philosophy with clear goals and priorities in keeping with the requirements of the socialist market economy and the need to contribute to the economic growth of the country.

Progress in implementing S&T reforms in agricultural R&D

Any overview of the impact of the new S&T policy on agricultural research in China has to acknowledge that although the organization and management of the system are similar throughout the country, quality and performance vary considerably. It is difficult to gain a comprehensive impression of any changes in the effectiveness and efficiency of the agricultural research service because of the size and complexity of the system, with different sources of funding and management at state, provincial, and county levels and the large number of research institutes and staff (Fan and Pardy 1992). For these reasons, the response to and effect of the policy reforms are not easy to measure.

The general impression of the impact of the reforms on agricultural research and technology was positive, and the staff of the research institutes interviewed at state and provincial levels agreed that the reforms had improved the research environment. This has occurred through the provision of new research opportunities, increased devolution of responsibility to the research institutes, and greater mobility of the staff, although this was largely restricted to the younger well-trained scientists.

Links with the market economy

There has been an increase in the interaction between the research staff and the end-users of technology, although this has not yet yielded sufficient money to permit any reduction in government appropriation of funds for the operational program of the institutes. One reason is that most research staff (with the exception of those involved at the prefectural and county levels, who deal with the application of research at the farm level) are not experienced in extension or in dealing with farmers at a commercial level. Those who are active and successful in their research resent the diversion, as it detracts from the time available for research, whereas those prepared to become involved in the transfer of research technology often receive little or no financial reward for their efforts.

Another reason is that many farmers and farm organizations are not prepared to pay or are not in a position to pay for technology. In some cases, where high-quality seed or propagation material of perennial crops, such as fruit trees, is offered, payment is less of a problem, but where the advice or technology is related to an activity regarded as public-good research, there is an expectation, from past experience, that it should be provided free of charge.

An exception to this generalization is the payment received by research institutes for assisting farmers involved in the TVEs based on agricultural enterprises under the Spark Program. Spark Program enterprises receive funding for specialized training and assistance with developing and using technology for the production or processing of agricultural products. At present, these funds are largely provided by grants from the state or the provinces, but in time this practice could become institutionalized as an integral service provided by the Spark Program corporations, and it could provide a valuable means of technology transfer for the research institutions and an important input in framing their research priorities.

The development of spin-off enterprises by agricultural research institutions has been quite restricted to date, in contrast to the use of this approach by the much larger group of engineering research institutes. At the same time, there is good evidence from the experience of the Shanghai Academy of Agricultural Sciences that this means of commercializing agricultural technology can be highly profitable. The Academy has several enterprises developed by its research institutes; however, the most successful is the Shanghai Mushroom Company, whose exports of edible fungi to Japan are worth 32 million CNY per year. Individual agricultural research institutes in other state and provincial academies in the country are only beginning to exploit this opportunity. Many of these have the information, improved plant or animal material, or special technology arising from their research that could be developed for both rural and urban markets.

Although the ability of the agricultural research institutes to participate in the market-orientated reforms to date has been limited, there is evidence that some of the actions being taken are having a positive outcome, including the exploitation of special products and some restructuring of research institute programs to focus on research topics that can support the research needs of the expanding rural TVE programs in their regions.

Other S&T policy initiatives

The funds provided for agricultural research projects are made available by SSTC through the respective ministries and other agencies that administer the scheme. These grants are made on a competitive basis and provide as much as 70% of the research funding for the national research institutes. The provincial academies also receive support from this source, in addition to grants from their own provincial sources. In some cases, agribusinesses also allocate funds for agricultural research, according to their own interests. In theory, these funds are allocated in accordance with the research program specified in the five-year plan, but as this is very general, the focus and nature of the research are largely determined by the ministries and academies involved.

Other research initiatives developed by SSTC over the last 10 years, such as the Torch Program (to assist in the development of NTEs), the National Key Laboratory Program, and NNSF grants for strategic (basic) research, are available to institutes engaged in agricultural research, but relatively few awards have been made. This is partly because of the applied nature of the research in agriculture, which may discourage applicants, and possibly because it is easier to obtain funds from the R&D grants administered by the Ministry of Agriculture or from provincial sources.

A recent evaluation of the National Program for Tackling Key Technology Problems in the Eighth Five-Year Plan (1990–95), a program administered by SSTC, was undertaken by the Chinese authorities (Government of China 1996). About 25% of the topics in this program are in agriculture, and the review provides some insights into the problems experienced in the management of research grants of this nature. The review commented on the difficulty of coordinating this type of program, which has too many levels of management and relies on other ministries and agencies to implement the program. Other problems involved reporting and serious delays in the provision of funds, which were severely reduced by inflation, poor management at the project level, and the lack of responsiveness of the program to the demands of the market. Moreover, many of the long-term projects lacked a clear focus within the major priorities of their particular areas of research. If these problems are common to other funding programs, such as the general grants that fund most of the agricultural research in the national academies (agriculture, forestry, and fisheries), it is urgent to address them.

Transfer of research technology

Spark Program

The Spark Program has been one of the most successful outcomes of the S&T policy reforms for agriculture. The program has now spread to virtually every province in the country and has helped develop 66 700 projects and many more individual enterprises within these. The sales from the projects in 1995 is estimated to be 260 billion CNY, and an additional 20 million people have found employment in rural areas. Possibly the greatest impact has been the increase in annual per capita income of the rural population in those areas where the Spark Program has been active. In a TVE visited in Jingyang County in Shaanxi, there had been almost a threefold increase in per capita income of the population of the county in the previous 5 years, and the target is to raise it to an average of 5 000 CNY per year by 2000.

The Spark Program is achieving one of the primary objectives of the agricultural policy, to stimulate and modernize the rural economy and improve the living standards of the farmers and their families. Further details of the program and its achievements over the first 8 years are given in the proceedings of a recent conference (SSTC 1994).

Observations by the review team, reports, and discussions with officials of the Spark Program suggest that a number of factors have contributed to the success of the program. The program is flexible — farmers can select from a wide range of well-developed technologies (projects) to suit their particular region or district — and is well linked to the local agricultural and industrial market systems. Consequently, the nature of the resulting TVEs in a given region or province varies considerably. In general, across the country there is an east to west increase in the number of enterprises that produce or process agricultural commodities.

Other features of the Spark Program that have contributed to its success are the following:

- The choice of a particular project within the program lies with the participants;
- The incentive to join the program is the prospect of a greater income;
- The technologies used in the Spark Program are in general already proven in practice;
- The selection of the leader of a Spark Program project is in the hands of the participants (subject to approval);
- Financial support is provided (from the state) for training the participants and for technical advice, usually from the local research institutes;
- The enterprises are funded almost entirely from bank loans and from capital raised by the participants and not from government grants, which tend to include more bureaucratic requirements; and
- A considerable effort is made to ensure that market outlets are available for the products of the enterprises.

One important outcome of the Spark Program is that it provides a focus for the agricultural research system in various parts of the country so that it reorients its research programs to service the needs of these rural clients. The program also provides the incentive to develop new and improved technologies to support existing enterprises and provides opportunities for new spin-off enterprises that have the ability to exploit niches in the evolving markets.

The response to these new opportunities that the Spark Program provides for the research institutions has been disappointing. Apart from the inputs mentioned above (technology and farmer training for the enterprises), there is little evidence that the research institutes are exploring new research opportunities in partnership with the more-advanced enterprises. Such partnerships would allow new technologies to be evaluated under realistic commercial conditions and, if successful, to be extended to similar enterprises. If the research institutes are to give effective leadership in S&T in the rural community, they will have to be more proactive, developing and evaluating new technologies before the demand arises.

Extension

There has been a major emphasis in the policy reforms on the transfer of research technology through a market-driven process. This works well in the industrial research sector, where technology is a merchantable commodity. In the natural sciences this approach has relevance and is being applied, but in general there are fewer opportunities. It will take longer to change the attitudes of the scientists and their clients toward selling and buying information and technology. This is particularly true in agriculture, where much of the research is of a public-good nature and the returns are reflected in the improved economy of the rural sector.

Another tradition in agriculture is that the transfer of research technology has been the responsibility of the extension service. Although this system has not, in general, been very successful, the system is in place in China and has a role to play in the new policy environment. The principal extension agency at the national level is the National Agricultural Technical Extension Centre, which is concerned largely with the formulation of government policies, whereas at the provincial level, the extension is organized under the departments of Agriculture and Forestry. In the past, institutional boundaries and fragmentation led to competition for resources and duplication of effort between research and extension, which was not productive. Another problem facing the extension service as a result of the market reforms and the decentralization of production units through the responsibility system is the need to service a very large number of individual farming families rather than the collective production teams of the former commune system.

To improve this situation, the state and provincial governments are funding large collaborative programs in which research and extension personnel undertake major development and extension projects, such as SEdC's Liaoyuan Program in Guangdong. The Bumper Harvest Program, developed by the Ministry of Agriculture, uses a similar approach and has been taken up by a number of the provincial academies of agricultural sciences.

Factors constraining the implementation of the S&T policy reforms

Research, extension, and funding of S&T for agriculture have consistently received the highest priority in the successive five-year plans and also in the S&T policy reforms, but the support received has never quite matched the rhetoric expressed in the policy statements. Part of the reason is that, despite the importance of agricultural research and the vital role it has to play in the future of China's rural development, the returns on investment in agricultural research are not as high as they could be, given the size of the research service and the number of research institutes and scientific staff. The reasons for this are discussed in the following section in conjunction with the three major goals for the S&T reforms in agriculture as set out in the 1995 Decision on Accelerating Scientific and Technological Progress. The diagnosis and suggested changes are drawn from experience with similar problems in agricultural research systems in Western industrialized countries.

Goal 1: Develop a modern, effective agricultural research service

Many of the problems lie in the complex structure of the agricultural research system, with research conducted in the research institutes of the Chinese Academy of Agricultural Sciences (CAAS) at the national level, in key national universities, and in the provincial academies of agricultural sciences, all of which operate independently. The lack of coordination resulting from this fragmentation (not to mention other agricultural research institutes, including those operating under CAS and ministries other than agriculture) leads to unnecessary duplication and competition for resources. Also, the disciplinary or commodity basis for research institutes, the lack of an output orientation, and the major research focus on the production aspects of agriculture do not fit with modern views on research organization and inhibit interdisciplinary interactions at the research level, which are essential for solving complex agricultural problems. The research service also largely excludes postproduction problems, including postharvest issues and those associated with processing the products of the production systems.

If China's agricultural research is to become a pacesetter in the world's advanced S&T, it will be necessary to overcome the constraints mentioned above and encourage greater integration of research and technology within the rural economy by reorganizing the structure and management of existing agricultural research and, through consolidation and integration, developing a more qualified service with a modern systems orientation.

The organization of research in the natural sciences in industrial countries has moved away from single-commodity or single-discipline research institutes toward a multidisciplinary model, located, where possible, in major agricultural ecoregions so as to address the special problems and opportunities presented by the commodities (plant, animal, or both), resources, and environments in the region. Such regional research centres are usually well equipped with a critical mass of scientists in the major relevant commodity and disciplinary areas. They are responsible for both the strategic and the applied research in the region. In China, collaboration of the national research institutes (CAAS, agricultural universities) with provincial research institutes located in the same region is essential to prevent duplication and to obtain assistance with provincial problems that form part of the regional research priorities.

In China, this model is one that might help with a number of problems that were discussed repeatedly during the Mission's visit, including

- Lack of interaction and the resulting duplication among research institutes and agricultural universities and colleges at national and provincial levels;
- The strong commodity or disciplinary focus at all levels of agricultural research; and
- The need for a more comprehensive set of research priorities, including new research to support the needs of farmers and agricultural TVEs in the different regions.

Goal 2: Strengthen scientific research and technology development

The earlier contributions of the agricultural research system that helped raise productivity to record levels in the post-1978 period came largely from the application of research solutions to what have been called first-order research problems, such as the introduction of improved crop and animal varieties, improved use and expansion of irrigation, the use of fertilizer, and the application of new chemicals to control pests and pathogens. The big gains from these inputs have now been achieved, and substantial further gains in China's agricultural production that will be needed to cope with a declining land area will require solutions to more complex, interactive, so-called second-order problems.

These second-order problems will require a more strategic (basic) multidisciplinary research approach, which in turn will require well-trained scientists with more sophisticated laboratory and field equipment and facilities. Some examples of these problems include the decline in factor productivity of the more intensively cultivated cropping soils, the need for better biological solutions to the problems of new pests and pathogens, and the search for improved quality and marketability of many of China's agricultural products.

With some notable exceptions, China's agricultural research has been slow to make the changes that began in the more industrial countries in the late 1960s. Providing advanced training for many people in China's agricultural research community will be a considerable task: as of 1990, only 0.2% of agricultural researchers had doctoral degrees. The agricultural research approach is changing; however, with such a large and extensive research system, the changes may not be fast enough to keep pace with the demands of the rural sector unless a much greater effort is made.

Once the research system is effectively restructured, attention should turn to the next most common factor constraining the delivery of research results, which is the management and performance of the staff of the individual research institutions. The criteria considered necessary for the development of an effective and creative research institute are

- A critical mass of high-quality and well-trained research scientists;
- Experienced support staff, with adequate equipment and facilities;
- Good leadership, delegation of authority, regular communication within the institute, and clear, well-focused goals and research priorities;
- Incentives in the form of more enlightened personnel policies, better living conditions, training, overseas travel opportunities, promotion on merit, and greater flexibility in employment; and
- Special funding to develop collaborative research activities for mutual benefit and to provide access to new skills, equipment, and facilities.

The S&T reforms in recent years have helped speed up the meeting of these criteria, but there is still some distance to go. Some of the most important requirements are the following:

- Strengthening the training and research skills of new young scientists, which should not be restricted to the graduates from agricultural universities, as many of the new skills required are from related disciplines;
- Providing modern equipment to permit a more in-depth approach to solving the more complex second-order (systems) problems that constrain agricultural production;
- Creating greater opportunity to undertake research-leadership training, to develop methods for setting priorities, and to open up new research in fields such as postharvest technology, food processing, biotechnology, resource management, development of livestock feed and fodder, and utilization of hilly land for wood, fruit, and animal production;
- Providing improved salaries and conditions to discourage the most able young scientists from moving to the private sector and out of strategic research — the quality of these people and their willingness to remain in publicly funded research positions long enough to develop successful technologies will be critical to meeting the growing demand for such research outputs; and
- Encouraging collaborative research as an important mechanism to overcome institutional and administrative barriers between complementary groups in different research institutes, university departments,

and industry — this has proven to be an efficient means of undertaking multidisciplinary research while ensuring the relevance and ease of transfer of the technology resulting from such research.

Goal 3: Improve the process of research and technology transfer to stimulate the economic development of the rural sector

Difficulties with the transfer of agricultural technologies are endemic in most countries and have not been solved by establishing large extension services, often in different ministries and usually separated from research institutions under a different form of administration. There has also been excessive focus on the means of transfer rather than on the transferability of the technology, which comes down essentially to the question of whether the technology can be readily adopted and be of benefit to the end-user and ultimately increase his or her profit margin.

The new S&T policies offer a new dimension for future extension activities in agriculture through the introduction of market-driven technology transfer, which de facto is becoming the norm in many developed countries. The adoption of this approach in the agricultural research institutes has been slow, but with the training of staff and the growth of agriculturally based TVEs, this approach will expand. One modification of the current structure of the research institutes that might help the process would be to group those staff members interested in extension, together with staff of the provincial extension services, in a special research division within the research institutes. This group's responsibility might include the design, evaluation, and extension of research technologies. Such a group would be closely associated with the research divisions but would be clearly responsible for evaluating the technology and moving it out to the end-users as rapidly as possible. To ensure equity in the sharing of the returns from the commercialization of research, a proportion of all funds derived from the sale of information and technology could be shared with the entire research staff of the institute as an incentive for both developing and transferring the technology.

In this respect the development of the new rural enterprises under the Spark Program provides a valuable new opportunity for agricultural research and extension activities. The Spark Program enterprises represent a useful framework against which the research programs can be better orientated and research priorities can be developed to service the current and likely future needs of the TVEs and the regional farmers in general. This, for example, would also develop a better balance between production and postproduction research activities, which is urgently needed to respond to the growth in the value-adding food-processing industries throughout China.

Research funding

The support for agricultural research at the national level is derived from SPC through SSTC to the ministries of Agriculture and Forestry, which administer the core funds for the research institutes of the academies of agricultural sciences and of forestry sciences, as well as the key agricultural universities. At the provincial level, the core funds for the equivalent institutions come largely from the provincial governments.

Research-project funds are allocated in accordance with the national priorities as set out in the five-year plan by SSTC in association with the respective ministries. National and provincial research institutes submit projects for funding on a competitive basis. The allocation is based on the quality of the research institute and the project and the funds available in the budget. Additional project funds are allocated to new research initiatives, including research of interest to agribusinesses. Part of the funds generated by institutes from the sale of their services or of produce can be used for research, but in most cases these funds represent a relatively minor component of the research budget.

The total funding for agricultural research in recent years has not been reduced. If anything, it has increased in real terms, especially at the provincial level; however, owing to the increase in total staff in recent years, the funds available per researcher have declined.

Because agricultural research institutes, along with research institutes in the other natural sciences, have a limited ability to profit from the market economy, they have not suffered from a loss of core funding, as has occurred in the engineering and industrial research institutes and in the physical sciences generally. Despite this, additional funds will be required to develop a high-quality agricultural research service with the capacity to undertake high-quality strategic and applied research, to provide advanced training, and to procure the equipment and facilities needed for modern research. One way to achieve this, although it would probably be difficult, would be to significantly reduce the large number of agricultural research scientists and research institutes, which would result in a major reduction in overhead, infrastructure, and operational costs. If these savings were spent in creating fewer but more comprehensive and better staffed and equipped laboratories, the current investment would be much more effective.

The imposition of levies on agricultural production is one potential source of additional funds for agricultural research. In many Western countries, levies charged on agricultural commodities and, in some cases, on manufactured products provide a significant source of revenue for agricultural research. The government negotiates the size of the levy with the industries and is responsible for collection at the point of sale. In many cases, the government also contributes funds on a one-for-one or some lower proportionate basis. The resultant research funds are allocated by an appropriate authority on the basis of research priorities jointly decided by the S&T authorities and the producers. This type of research tax on production is especially appropriate for many agricultural commodities, such as rice, wheat, and livestock, that are the beneficiaries of public-good research.

Chapter 10

ENVIRONMENTAL AND SOCIAL DEVELOPMENT

S&T reform policies and environmental development

During the decade of S&T reforms, 1985–95, two related trends appear to have taken place in China's environmental development: the combined effects of rapid industrialization, increasing agricultural production, and growing population pressure caused environmental degradation in rural and urban areas; and the attention paid to environmental S&T increased. What happens to China's environment is significant at the global scale. The decade of S&T reforms has generally reinforced the importance accorded by the government to environmental R&D for China's economic and social development. After the Stockholm Conference on the Environment in 1972 and particularly after the United Nations Conference on Environment and Development in 1992, Chinese leaders began to give more attention to the problems of environmental-quality degradation and natural-resource depletion, and this has been accompanied by supportive S&T policies.

The key processes of the S&T reforms affecting the environmental sector include the following:

- Stimulating institutional reform through declining government budget allocations;
- Creating a market for environmental research and technology;
- Setting strategic research priorities;
- Establishing the Spark Program for rural development, which has created thousands of TVEs and many new point sources of pollution;
- Establishing the Torch Program for NTEs;
- Opening up of China to international S&T forces; and
- Establishing China's Agenda 21 program for sustainable development.

Progress and difficulties in meeting S&T reform goals

The Mission visit, both the site visits and discussions with Chinese experts and officials, provided the following impressions of progress and difficulties in meeting the S&T reform goals, through the prism of environmental development.

Stimulating institutional reform by introducing competitive (or competing) government budget allocations

The government's decision to force S&T institutions to respond to the market by cutting their operational budgets was implemented over the period 1986–90. Three main types of institutions were distinguished for differential cutting, depending mainly on their assumed ability to market their goods and services: technology-development institutions, basic-research institutions, and public-welfare S&T institutions. Both agricultural and environmental-protection S&T institutions were classified as undertaking public-good research, and their structure and organizational cultures were relatively protected from the impact of the reform policy. The budget breakdown for 1991, at the end of the implementation period, shows that environmental S&T institutions continued to receive 72% of their budget from the national government (Table 13).

New environmental institutions were created during the period of budget reforms: these include the CAS Institute on the Environment and the Committee of Environmental Protection and Natural Resource

Table 13. Sources of income for S&T institutions, 1991.

	Government	Market	Bank loans	Other
Industrial technology (%)	22	61	12	4
Agriculture (%)	55	33	5	6
CAS (%)	68	21	1	10
Environment (%)	72	27	2	<1

Source: SSTC (1992).
Note: CAS, Chinese Academy of Sciences; S&T, science and technology.

Conservation of the National People's Congress, which was created in 1992 to oversee the development and implementation of environmental legislation. On the Mission's visits to environmental S&T institutions, it was told that opportunities for earning income by selling environmental technologies are very limited. However, experience from other countries suggests more opportunities for exploration in China — for example, in the areas of environmental consulting, environmental-impact assessment, and environmental-risk management. In addition, and more telling, the Shanghai Light Industry Research Institute (see case study, Box 5) has demonstrated that niche markets are to be found for any institution with the drive to identify them. These are likely to be growth areas in China within the next few years and are fields in which Chinese skills should be developed, instead of relying on foreign consultants.

Within environmental S&T institutions, restructuring to reduce the number of departments, to identify the core (tenured) personnel, and to facilitate multidisciplinary project teams is taking place, similar to that found throughout the S&T system. We saw less evidence that effective horizontal linkages were being created between institutions, either between different CAS institutions working on environmental issues or between CAS, CAAS, and ministry institutions that are working on closely related issues and would benefit from greater interaction — for example, in the fields of sustainable agriculture and land and water management. Some coordinating institutions exist at provincial and local levels, but these institutions indicated that their ability to coordinate the work of national S&T institutions located in their jurisdiction was limited.

The research system in China, as in other countries, is generally structured along disciplinary lines, so interdisciplinary research has to overcome organizational boundaries, both within and between institutions. For environmental research, this is a major challenge because environmental problems require the inputs of many disciplines, both in building models and in empirical research. Research institutions at national and provincial levels are restructuring, and this may facilitate more interdisciplinary research. The establishment of strategic research grants for interdisciplinary environmental projects will further these reforms. The current rethinking of the structure and role of CAS might also take the particular needs of environmental research into account and strengthen the linkages not only between the various environmental sciences but also between these sciences and policy research.

The Mission got the impression that for the environmental S&T sector, there was more progress to be made in improving the coordination between research institutions and between these institutions and the production sector. This further rationalization of the environmental S&T system could eliminate duplication of research efforts and better meet the research needs of the different regions of China by responding not only to national priorities but also to those priorities defined by provincial S&T authorities. Thus, although we were impressed by the work and achievements of many individual institutions, we were less convinced that the environmental S&T institutional system was as cost-effective and innovative as it could be.

Creating a market for environmental research and technology

Environmental S&T institutions cover a wide spectrum, from basic research on Earth sciences and ecology to pollution-control technology. They do not form a coherent group, in terms of either disciplinary backgrounds or orientation to the market. The Chinese government has identified basic environmental sciences as a strategic national priority for research in relation to both sustainable food security and global environmental change. These are the parts of the environmental S&T system that the government is prepared to "anchor"; other components, such as pollution-control technology, are encouraged to enter the marketplace.

Particularly important for China's development, among the market-oriented components of the environmental S&T system, is innovation in developing a domestic environmental-protection and -monitoring technology-production sector; cleaner production technologies for manufacturing; and clean energy production, especially washed-coal technology. In each of these areas, China is faced with major and urgent environmental impacts, a domestic capability that lags behind the capabilities of industrialized countries, and a desire to become technologically advanced. International concern about China's rising contribution to global change and Japan's self-interest in reducing the acid precipitation believed to originate in China have led to international assistance and joint ventures for the necessary capital to help develop these sectors.

The first step in the planned technology-transfer process is to import advanced equipment and expertise (including not only technology but also management systems and quality-certification and information services), mainly through joint ventures. The next step is to build up a domestic environmental-protection and -monitoring technology-production sector. The task is formidable, as it involves the debt-burdened and inefficient SOEs and the small and widely dispersed TVEs. China currently has more than 1 800 enterprises producing environmental-protection equipment; 80% of these are TVEs, producing products of various qualities, and the TVEs are probably themselves polluting the environment as they manufacture the equipment.

There is also a potentially very large market for environmental-information services in China and overseas, and this opportunity seems to the Mission to have been less well explored by the environmental S&T institutions. The Chinese government has already allocated 3 million USD, plus annual maintenance costs, for the establishment of the National Climate Centre, which will bring together research in climate modeling and scenario analysis, experience in response strategies, and long-term geological and historical data. Few countries have such a richness of historical record to combine with advanced modeling and geographic information system data. Environmental information and analytical research also have direct application to land-use zoning applications, environmental-impact assessments for infrastructure projects (such as the major ones planned for water diversion and transportation), and the expanding market for environmental-risk assessment. The government would itself be a major consumer of such services, and there is a need for government support for basic research, but it may be worth exploring how far such a national centre and similar environmental-information sources could generate some revenue by selling their services to a market that would include the large number of new foreign investors in China's developing economy.

Environmental scientists and engineers have benefited from the policy reforms to enhance the status of intellectuals and S&T personnel in society, but the Mission did not see much evidence of large numbers of environmental scientists leaving the S&T institutions to join or to start enterprises — rather the reverse. Despite declining institutional budgets, we were told only a few scientists were leaving the environmental S&T institutions; more were taking a second job, often in environmental consulting. The Mission was unable to assess the degree of mobility of environmental scientists, which is likely to vary widely among scientific disciplines and institute locations. We believe that the socialist market economy presents more opportunities for mobility than seem to be taken advantage of at present.

Setting strategic research priorities

The Mission was impressed with the degree to which China has articulated its strategic research priorities for the medium and longer term. The establishment of a competitive allocation process for basic and applied research through the NNSF of China in 1986 led to priority setting, even within the part of the S&T system that was to remain anchored to government support. Overall, environmental research ranks high among the priorities within the general, key, and major programs of NNSF funding. Environmental research areas given priority for funding include clean energy sources; energy-saving processes; environmental chemistry; pollution control; a broad number of priorities within Earth, atmospheric, and ocean sciences; and a major strategic program on global-change research.

This commitment to support basic research in Earth and environmental sciences is important, not only to China, but internationally, as global-change research now ranks as big science or involves, in OECD's terms, megaprojects. It requires large amounts of funding and international coordination of national priorities for research funding because the current generation of global models of Earth–ocean–atmosphere interactions requires comparable data inputs from around the world. China is playing a leading role in global-change research and the associated international research programs. The S&T reforms have thus strengthened China's contribution to international research in global change.

Establishing the Spark Program for rural development

The impression gained by the Mission is that, partly as a result of the S&T reforms — particularly the establishment of thousands of TVEs throughout the rural areas, largely through the Spark Program — the environmental challenge to the S&T system is considerably greater now than at the beginning of the decade of reform. The environmental policy response will thus have to be correspondingly more innovative to be effective. The TVEs include many that are highly polluting in the traditional sense, such as coal mining, cement making, pulp and paper production, and brick making. They also include new industries, such as pharmaceuticals, fertilizers, and plastics, that create waste that is less visible but that is more difficult to monitor and can be locally very harmful to human health and to the environment.

The TVEs have been very successful in absorbing rural labour and increasing rural incomes. In 1994, TVEs accounted for about 75% of the rural GDP and employed 28% of the rural labour force, or more than 120 million people. They are the source of 65% of the net income of Chinese farmers and now employ more people than the SOEs. By 2000,

TVEs are expected to account for 45% of China's GDP and 80% of the rural GDP. They are expected to grow especially rapidly in the central and western inland regions, where there are preferential policies.

TVEs are thus a major force in national development and in income generation for the rural areas. They are seen as growth poles for new towns and industrial zones and a mechanism for regional equalization, through a program of east–west cooperation and management training. These TVEs are likely to create thousands of point sources of pollution, affecting air, water (surface and groundwater) and land, unless more emphasis is given both to strengthening environmental monitoring and to ensuring that the technology packages provided through the Spark Program are clean production technologies. As has been the experience of SOEs, retrofitting of old, polluting technology is expensive and often neither technically nor economically feasible. The cost of trying to clean up millions of hectares of agricultural land and the associated water bodies would seriously affect China's economic growth and is best avoided by accelerating the process of implementing environmentally sound policies for TVE development as a matter of urgency.

On the basis of what we heard and saw, the Mission also felt that the program of environmental monitoring and regulation for TVEs could be improved and that perhaps here was an opportunity for more cooperation between the scientific institutions (some of which have analytical capability for measuring pollutants) and local EPAs (which generally do not).

More broadly, the approach to environmental regulation in China as in other countries is shifting from downstream or end-of-the-pipe control to upstream anticipatory and preventive policy instruments. The difficulty of the current emphasis on the polluter-pays regulatory principle (even including the death penalty) is that many of the TVEs, once established, have little economic or technical capacity to adopt cleaner production technology, to pay significant fines, or even to monitor their own emissions. Thus, in rural industrialization policy, the S&T reforms have created an environmental moving target for themselves, necessitating further reform in S&T priorities, system renewal, and performance. This is not unique to China, although the speed and scale of rural industrialization in China probably are. The challenge is to find an appropriate Chinese package of S&T initiatives to deal with this second-generation environmental problem.

Establishing China's Agenda 21 program for sustainable development

The various initiatives relating to environmental protection and natural-resource conservation that featured in earlier S&T reform programs are brought together in *China's Agenda 21* (SPC and SSTC 1994). This comprehensive strategy document will function as a guideline for drawing up medium- and long-term plans for economic and social development, and its goals and programs will be included in both the Ninth Five-Year Plan (1996–2000) and the plan for 2010.

The White Paper on Agenda 21 was prepared by a leading group co-chaired by a deputy minister of SSTC and a deputy minister of SPC, who coordinated the input of 52 ministries and agencies and more than 300 experts during 1993. China's Agenda 21 was adopted at the 16th Executive Meeting of the State Council on 25 March 1994. It includes 78 program areas and their goals and objectives; of these, 9 priority programs have been identified for the first round of funding and action. It is anticipated that 60% of the funding for implementation will come from China and 40% from international assistance, cooperation, and investment. It was reported during the Mission visit that already 3 billion USD was committed from international sources, primarily the World Bank, the Asian Development Bank, and Japan.

The nine priorities for the first round of the Agenda 21 program are the following:

1. Capacity-building for sustainable development;

2. Sustainable agriculture;

3. Cleaner production and an environmental-protection industry;

4. Clean energy and transportation;

5. Conservation and sustainable use of natural resources;

6. Environmental pollution control;

7. Poverty and regional development;

8. Population, health, and human settlements; and

9. Global change and the Convention on Biodiversity.

The preparation and approval of the White Paper on Agenda 21 and the establishment of a coordinating committee to oversee its implementation are evidence of the importance that the Chinese government, at the highest levels, attaches to making environmental protection an integral part of economic and social development. The White Paper is a major achievement in integrated national planning for sustainable development and is a model for other countries. The Mission was also impressed with the awareness that we encountered among people at all levels, not only of the importance of maintaining environmental quality and of the main environmental problems in their area, but also of the Agenda 21 program itself. The coordination achieved in setting out the priorities and projects in Agenda 21 shows what can be achieved across the S&T system in terms of collaboration and multidisciplinary research focused on economic and social development.

Deepening the S&T reform process for the environmental sector

The Agenda 21 program anticipates a further development of the legal, policy, and educational bases for sustainable development, and in the view of the Mission, this is a necessary corollary to the current priority accorded to environmental S&T in China.

During the 1980s, a number of key environmental-protection and natural-resource-protection laws, regulations, and standards were enacted. These rely heavily on administrative regulation, including the ability to monitor environmental quality at the local level, and embody the polluter-pays principle. The Chinese government has already identified where there are short-term needs to fill in specific gaps in the current legislation. Among the important new legislation being developed is that for solid-waste pollution and for the environmental management of TVEs. More important, the national government recognizes that in the longer term, the legislative and policy framework has to evolve toward policy instruments that are more consistent with a market economy. These include specific environmental taxes, pollution levies, tradable permits, and economic incentives. This is generally uncharted territory for any government and particularly for China's, with its unique task of developing a socialist market economy.

Important elements in this policy package include the following:

— Integration of environmental protection and rational resource use into other policies, especially industrial policies;

— Policy feedback mechanisms for assessing the environmental impact of other economic, social, and S&T reforms and adjusting them appropriately;

— Basic research in environmental systems and policy analysis to understand both the ecosystem and policy-system behaviour and their linkages; and

— An upstream policy emphasis, that is, relying more on prevention than on cure.

Part III

Account of Final Meetings in Beijing, Shenyang, Xi'an, and Shanghai

Chapter 11

REVIEW OF CHINA'S S&T POLICIES

The international review team submitted Parts I and II of its report to SSTC at the end of December 1995. The report was translated into Chinese and distributed by SSTC to most of the organizations interviewed during the Mission's earlier visit to China. The Mission team then returned to China in May 1996 for a series of discussions and debates on Parts I and II of the report. This section, Part III of the report, summarizes these discussions, which occurred in Beijing, Shenyang, Xi'an, and Shanghai.

Discussions in Beijing

The meeting in Beijing was held on 20 and 21 May. It was convened by SSTC and chaired by Madam Zhu Li-lan, Executive Vice Minister of SSTC. More than 100 representatives of government commissions and ministries, institutions of higher education, and business enterprises participated in the event. It was one of the first occasions on which SSTC had brought together such a wide cross-section of stakeholders concerned with S&T to debate the S&T policy reforms.

Sessions were held on eight topics identified by SSTC. Introductory comments were made for each topic by a member of the Mission and a member of SSTC. There was then a general discussion of the issues raised.

The agenda for the Beijing meeting was as follows:

Opening Session
    Mme Zhu Li-lan, Vice Minister, SSTC
    Dr Geoffrey Oldham, Mission Leader

Topic 1 — Overall Impressions of the Reforms
    Dr Geoffrey Oldham
    Mr Huang Yingda, Deputy Director General, Department
      of S&T Policy, Legislation and System Reform, SSTC

Topic 2 — Reform and Development
    Dr R. Peter Suttmeier
    Mr Zhao Yuhai, Deputy Director General,
      Department of Planning, SSTC

Topic 3 — National System of Innovation
    Mr James Mullin
    Mr Shi Dinghuan, Director General,
      Department of Industrial S&T, SSTC

Topic 4 — International Collaboration
    Dr Geoffrey Oldham and Dr Greg Tegart
    Mr Jin Xiaoming, Deputy Director General,
      Department of International Co-operation, SSTC

Topic 5 — Sustainable Development
    Dr Anne Whyte
    Mr Chen Yuxiang, Deputy Director General,
      China Agenda 21 Management Centre, SSTC-SPC

Topic 6 — Revitalising Agriculture with S&T
    Dr James McWilliam
    Dr Wang Hongguang, Director, Division of Policy,
      Department of Agricultural S&T, SSTC

Topic 7 — Basic Research
    Mr James Mullin
    Mr Wu Shuyao, Director General,
      Department of Policy, NNSF

Topic 8 — High Technology
    Dr Greg Tegart
    Mr Shao Liqin, Director General,
      High Technology R&D Centre, SSTC

The Mission members summarized the main findings of the Mission on each issue, as documented in Parts I and II of the report, and the SSTC representatives presented the highlights of the reform policies and indicated the steps being taken to address recognized shortcomings and weaknesses. In the debate that followed these presentations, there was very little disagreement with the views and conclusions reached by the Mission. Further examples were given of experiences that tended to confirm these conclusions. Nevertheless, the inherent weaknesses of the approach followed in the review process were recognized by all participants. China is a huge country, encompassing a great variety of experiences. It is difficult to generalize from discussions with only a few hundred individuals and institutions. Despite these limitations, there was general agreement that the opportunity the review had provided for a mixed group of Chinese stakeholders to debate the S&T policy reforms in an international context was very worthwhile.

The following is a brief summary of the main points made at each of the sessions. A more complete summary of the meetings will be published by the SSTC.

Opening session

Madam Zhu Li-lan opened the meeting by recollecting the important changes in Chinese S&T since the March 1985 Decision on the Reform of the Science and Technology Management System. These changes include the following:

- Progress in the move away from the previous unitary and closed system of planning, which had divorced S&T from the economy;

- Opportunities for the development of nongovernmental S&T enterprises and of new and high-technology industrial zones;

- Focus of R&D resources at the central, provincial, and municipal levels on application to economic development, the expansion of new and high-technology industries, and the strengthening of fundamental research — this has been accompanied by an expansion of S&T education;

- New steps in legislation on S&T;

- Emergence of a "soft-science" research community, which has made the process of decision-making more democratic and more scientific — this community has become an important adviser to governments at all levels; and

- Increasing international cooperation and exchanges in all fields of S&T as China's economy and its S&T become much fuller participants in the international system.

The present context for further reform of S&T in China is marked by

- Continuing improvements in the policy environment that favour the integration of S&T within the Chinese economy;

- An urgent need to promote economic development based on advances in S&T;

- A need for a greatly increased innovative capability as China enters the international economic system; and

- A need to accelerate the development of S&T and innovation and to incubate and foster the growth of new and high-technology industries to catch up with global trends.

The review of China's reform policies for S&T has been based on interviews in China with many decision-makers in the S&T system, R&D managers, science policy researchers, and representatives of research institutions, universities, SOEs, start-up companies, and TVEs. During the work of the review, valuable contributions were made by CAS, CAST, NNSF, SEdC, the Ministry of Agriculture, several provincial and municipal STCs, and many universities and enterprises. SSTC wishes to offer sincere thanks to all of those who contributed to the work so far.

In his response to Mme Zhu, Dr Geoffrey Oldham indicated that the Mission had used the approach pioneered by OECD for reviews of national S&T policies. The International Development Research Centre (IDRC) had previously sponsored a similar review in South Africa, and the United Nations Commission on Science and Technology and the United Nations Conference on Trade and Development are extending this approach to developing countries. Hence, the approach followed in China was built on a well-tried methodology.

The approach consisted of three stages. The first stage involved the assembly of background documentation. The second stage consisted of a 3-week visit to China by the Mission, in November 1995, during which interviews took place in Beijing, Shenyang, Xi'an, Shanghai, and Guangzhou. The third stage consisted of the final meetings, held in Beijing, Shenyang, Xi'an, and Shanghai in May 1996.

The Mission is all too well aware of the limitations of this approach. The visit to China was for only 3 weeks, and although several hundred people were interviewed, either individually or in small groups, there are whole areas of Chinese science, such as the medical sciences, from which no representatives were met.

It should be noted that the Mission's report makes very few recommendations. This follows the usual approach for this type of review. The intentions have been to

- Reflect back to China what the Mission has been told — in the available time, it was not possible to carry out independent reviews of any Chinese S&T institution; and

- Put what was learned into an international context — the aim was to try to identify those areas in which international experience may have some relevance for China, so when the report does make recommendations, it is mainly to suggest some specific international experiences that appear to be worthy of detailed study by China.

Finally, Dr Oldham indicated that the Mission wished to acknowledge the tremendous support and encouragement received from SSTC. All Mission members were also grateful to the hundreds of people throughout China who answered many questions with great patience and who helped form the Mission's impressions of what must be one of the most profound changes in the organization and mobilization of S&T in the history of any country.

Topic 1 — Overall Impressions of the Reforms

Mission comments

Four main impressions were highlighted:

- Much care and thought had been devoted to developing S&T policies that contributed to the goal of creating a socialist market economy;

- Implementation of the policies seemed to be variable — where managers were still in the mind-set of a command economy, implementation had been slower than where managers had embraced reform and were prepared to innovate;

- There appears to be a rigid demarcation of responsibilities among commissions and ministries at all levels of government, leading to a plethora of policies, schemes, and initiatives with little coordination; and

- A number of S&T policy issues that are considered important in other countries have been relatively neglected in China — these include mechanisms for providing independent S&T policy advice, an NSI approach to policymaking, and the use of foresight techniques in priority setting for strategic research.

Chinese comments

In the discussion that followed, it was considered that two of the main accomplishments of the reforms had been to change the financing system of S&T activities and to create a technology market. Among continuing problems in need of further attention, the following were identified by the Chinese participants:

- The continued existence of too many S&T institutes dependent on government financing;
- The need to clarify IPR issues; and
- The need to clarify ownership rights to technology, particularly in the case of technology generated in institutes and laboratories no longer financed by the state.

Topic 2 — Reform and Development

Mission comments

There has undoubtedly been great progress in S&T developments as a result of the overall reform process. Many of the issues noted by the Mission have already been addressed by the May 1995 Decision on Accelerating Scientific and Technological Progress. But this Decision has important implications for

- Research-production linkages;
- Relationships between the importation of foreign technology and domestic innovation;
- The possibilities of reaching the 1.5% GERD-GDP target; and
- Ways to achieve better coordination.

There is a need for even more institutional changes if the full potential of the reforms is to be realized. The relationships between institutions within central government and between central and local governments will need to be studied and clarified if government efforts to promote innovation are to succeed.

Technological innovation is primarily a function of the behaviour of enterprises. There are now a variety of enterprise types: SOEs, TVEs, NTEs, joint-venture enterprises, and private and foreign enterprises. Most of these did not exist before the reforms. How do these enterprises compete and cooperate? And how can governments use this new industrial pluralism to promote science in the industrial sector?

Reform and development are dynamic processes that can solve problems, but they also create new ones. Policymakers need access to knowledge about the changing conditions from the stakeholders who are most involved. That is why the Mission has drawn attention to the need for new S&T policy advisory mechanisms.

Chinese comments

The way in which SSTC is responding to the need for new policy initiatives was demonstrated by a description of the role of S&T in the Ninth Five-Year Plan.

The general goals for S&T in the plan are

- To serve economic progress by promoting innovations to improve labour productivity and economic competitiveness;

- To improve the quality of life through more attention to sustainable development; and

- To serve national strategic goals for S&T in the 21st century, allow Chinese S&T to take its place in the international scientific community, and make a contribution to the knowledge stock of humankind.

Agricultural S&T is to be given the highest priority in the plan, in recognition of the economic importance of agriculture and its relative backwardness in the face of considerable challenges posed by population growth and diminishing available land.

Other priorities are high-technology development, S&T in support of Agenda 21 objectives, and basic research. High-technology products will come to occupy a larger share of China's overall industrial production (from 10% to 20%) and of manufactured exports (from 7.9% to 15%) by the end of the plan period.

The achievement of the plan's objectives will require improved coordination of government agencies, better management of R&D (and innovation generally), and more funding.

There will be new attention given to interagency coordination and to the development of policies that are sensitive to market functions. In response to Mission concerns about reaching the 1.5% GERD–GNP target, policies to encourage enterprise spending on R&D are being developed and local governments will also be spending considerably more on R&D from local sources.

Topic 3 — National System of Innovation

Mission comments

In industrialized countries, and in a growing number of newly industrializing countries, policymakers have found that the concept of an NSI provides a useful framework for technology policy formulation because it makes explicit the many different kinds of inputs needed to produce an economy that is innovative and hence competitive in today's increasingly globalized markets.

The Mission's report sets out a series of six sets of functions (see Table 2) that need to be present in an effective NSI:

Central government functions	Shared functions
Policy formulation and resource allocation at the national level	Performance-level financing of innovation-related activities
Regulatory policy-making	Performance of innovation-related activities
	HRD and capacity-building
	Provision of infrastructure

The use of an NSI as a framework for policy signals a radical departure from the current situation and understanding, replacing the old with the new view of the role and status of S&T and engineering in national development. The perception by many countries that technical change is the primary source of economic growth means that economic and S&T policies have to recognize as central concerns the two processes — innovation and technology diffusion — that are the agents driving that technical change.

The four key interests of any country can be thought of as being to ensure that

— The country has in place a set of institutions, organizations, and policies that give effect to the various functions of an NSI;

— There are constructive interactions among those institutions, organizations, and policies;

— There is in place an agreed-upon set of goals and objectives that are consonant with an articulated vision of the future that is being sought; and

— There is in place a policy environment designed to promote innovation.

As an indicator of how the idea of an NSI can be used in policy analysis, the Mission report uses a series of tables (see Tables 4–8) to highlight the different roles of different participating institutions or stakeholders in China's NSI.

Chinese comments

The notion of a Chinese NSI was considered a useful one, and a number of studies on the topic have already been carried out. Current studies involving SSTC and other commissions and ministries are dealing with a strategy to promote industrial technological innovation and to develop a Chinese NSI under the conditions of the development of a socialist market economy.

The present innovation situation in China was characterized as having the following features:

- There are great opportunities in increasingly open markets, both domestically and internationally, and considerable challenges posed by increasing globalization of competition and problems concerning intellectual property and environmental protection;

- The innovative capacity of much of Chinese industry is weak — much of the technology used is 15–20 years behind the times, rates of commercialization of new technology are slow, and only a small minority of enterprises (14% in one survey) make any attempt to upgrade international technology after it has been imported;

- Chinese enterprises underinvest in R&D compared with their international competitors, their management is unfamiliar with the promotion of innovation or the management of technical change, institutes that are not part of enterprises have great difficulty establishing close linkages with enterprises, and China lacks an effective system of technology diffusion; and

- If China is to use market mechanisms to stimulate innovation, then there is a need to develop an appropriate macrolevel management system and policy environment that are adapted to a market economy.

To overcome these problems and build an NSI, it is proposed that China take action along the following lines:

1. Establish an institutional framework for promoting innovation within which firms are the main force, in which links between industry, institutes, and universities are strengthened, and in which innovation is seen as the mechanism driving technological progress.

2. Establish an innovation-driven policy and regulatory system.

3. Establish the government's macroadministration system, with a clear delineation of the boundary between government and enterprises.

China's final goal is the establishment of a market-compatible NSI. China has embarked on a series of pilot experiments in some key industries to explore means of promoting "innovation engineering" and an NSI.

Topic 4 — International Collaboration

Mission comments

The Mission made a distinction between collaboration in science and collaboration in technology. With regard to science, national governments have different motivations for promoting collaboration. It is important to recognize these different objectives and to develop strategies to maximize the national benefits of collaboration and minimize the disadvantages.

China produces only a small fraction of the world's R&D and therefore needs to access the world's store of knowledge. International S&T collaboration is one way of achieving this objective. China has recognized the value of collaboration and has initiated many collaborative programs. The Mission felt that, perhaps, there needed to be a more focused approach to collaboration.

Some of the problems identified in discussions in China were

– Access to foreign journals;

– Widespread access to the Internet;

– IPR; and

– Participation in big-science projects.

Chinese comments

China does have a policy for international S&T cooperation. The principles that underpin the policy are

– Country specificity (SSTC has developed a series of individual-country policies, which are updated every 2 or 3 years);

– Equality and reciprocity;

– Complementarity;

– Protection of IPR;

– Contribution of joint activities to S&T development and economic growth; and

– Contribution of joint activities to China's foreign policy (in certain cases S&T cooperation can be a forerunner of larger foreign-policy initiatives).

Over the last 18 years, China has established cooperative S&T relations with 145 countries, has signed formal agreements with 96 of them, and has joined 875 international S&T organizations.

Some new challenges to policy formulation for international S&T collaboration, created by rapidly changing global circumstances, include

- The need for a policy response to multinational corporations now establishing R&D facilities in China (China has a positive attitude to this development but has yet to put in place some necessary legislation);
- The need to consider the expansion of access to the Internet by Chinese researchers; and
- The need to improve the enforcement of Chinese laws and regulations pertaining to IPR protection (including China's own IPR).

China is formulating a new international strategy for S&T collaboration that will address a series of objectives in the context of increasing globalization; future cooperative activities will be expected to

- Contribute to the country's national security and overall foreign-policy objectives;
- Serve the country's economic growth and S&T development; and
- Promote sustainable development.

Within those broad objectives, some of the new themes that will be pursued are

- Strengthening the links between S&T cooperation and foreign trade (China wishes to increase the volume of high-technology products and services that it can offer on international markets);
- Increasing the extent of cooperation in the development of industrial technology (for example, in advanced manufacturing technology);
- Improving the assimilation of imported advanced technologies;
- Assisting R&D institutions in their efforts to establish branches overseas;
- Creating a national fund for international S&T cooperation; and
- Exploring effective channels for China's participation in global big-science projects, including industrial technology cooperation within those projects.

Topic 5 — Sustainable Development

Mission comments

The Mission was impressed with China's commitment to the goal of sustainable development and with the efforts, including China's Agenda 21, being made to integrate environmental protection into economic and social development.

A distinction can be made between the environmental S&T sector, on the one hand, and the impacts on environmental quality and the threats to sustainable development that come from other sectors of economic development, on the other.

Environmental technology, together with the scientific management of natural resources, has the potential to be a major growth sector in the national economy. The major resource-development and infrastructure projects in China present opportunities for China to develop its own capacity in environmental management and assessment instead of relying on foreign technology and expertise.

The Mission heard from many people that China's environmental quality is, in general, getting worse. This deterioration of the environment is taking place despite great efforts on the part of the government and despite having protection of the environment as a basic national policy in China. Sustainable development is the most difficult test of the overall success of the S&T reforms, and it raises particular challenges to implementing the reforms.

To tackle these problems, the Mission felt, will require a greater effort to integrate policies, both those of different levels of government and those of departments and agencies at the same level.

The implementation of S&T reforms has occurred more rapidly than that of other policy reforms, with the result that future S&T initiative reforms now depend on wider policy change. For example, the Mission learned that the prices of resource inputs, such as energy and water, are much lower than the total costs associated with them (including indirect environmental costs and health costs) and are even lower than the direct economic costs of supply. This leads to inefficiencies, overproduction of waste, and increased pollution. Thus, low prices for coal and other sources of energy are actually acting as perverse incentives to pollute, and low prices for water are acting as incentives to waste it. The problem is a lack of policy integration at the national level across economic development sectors.

A Green Spark Program? — S&T reforms have sometimes created problems for sustainable development when wider policy integration is not achieved. One example of this is the TVEs, which include heavily polluting industries, such as mining, metal processing, brick and cement making, asbestos processing, and pulp and paper making. In creating the TVEs, China has increased the incomes of the rural population but has also established thousands of small point sources of pollution, which are widely dispersed and undercapitalized, use outdated technology, and have generally low technical and managerial skills in the work force.

The Mission felt that the very characteristics that make TVEs such a formidable challenge for environmental monitoring and control may provide an opportunity for a new SSTC-initiated program that would build on the successful characteristics of the Spark Program but would provide the TVEs with proven packages of technology and expertise to achieve low-cost, cleaner production systems.

Chinese comments

The objectives and early implementation of China's Agenda 21 program were elaborated and given as an example of coordinated government action. It was agreed that different sectoral policies can be compatible at the national level but can run into conflict when implemented at the local level, where specific targets and actions make short-term inconsistencies between national and local policies or between ministerial policies more explicit.

The issue of economic incentives and penalties to control pollution was raised, along with the question of how a national policy could arrive at an appropriate balance between them. In addition to having both incentives and penalties, it was pointed out that policy frameworks should include both voluntary and mandatory incentives. China should move toward full cost accounting for major products and services, beginning with major polluting industries. The full costs should be better reflected in prices and should include not only environmental costs, in terms of pollution and resource depletion, but also health costs to workers and to nearby communities. International trade agreements are likely to pay increasing attention to full cost accounting and will enforce trade sanctions against countries that achieve low export prices for their products by not including the environmental and health costs of production. The North American Free Trade Agreement side-agreements on labour and environment were given as examples of a sustainable-development approach to export prices being embodied in multilateral legislation.

The recent concern in China about receiving foreign waste for recycling without adequate controls on inspection at the port of entry led to a discussion on the undesirability of China's becoming a pollution haven for multinational firms to set up production using outdated technology and high levels of pollution emissions. Maintaining high national environmental standards, together with adequate monitoring and enforcement capability, was emphasized as being the best way for China to achieve sustainable development and attract industry with the best international practices.

Topic 6 — Revitalising Agriculture with S&T

Mission comments

Obtaining an accurate measure of the impact of the S&T policy reforms on the 1 200 or so agricultural research institutes (including agricultural universities) above the county level is difficult because of the size of the system and the complexity of its management and its funding sources. However, the Mission's impressions are that the policy changes have improved the research environment by providing new research opportunities and funding and giving greater responsibility to research directors for the management of staff and research programs.

Despite this, the ability of the great majority of research institutes to generate additional funds by exploiting the market economy has been limited, as is true for much research in the natural sciences. One of the reasons for this has been the failure in some cases to change the organization, management, and funding of the R&D system to match the new market-driven research environment. In addition, staff in the R&D system generally lack experience in operating in the commercial sector; research-based technology is not always appropriate to the needs of that sector; and the clients (end-users of the technology) are reluctant or unable to pay for much of the technology derived from research of a public-good nature that, in the past, had been provided free of charge.

Technological innovation through research is a major catalyst in development and economic growth, and hence the creation of a more vital and effective agricultural research service is crucial if the goals of contributing to the economic and social development of the rural community are to be achieved.

The revitalization of the R&D sector is constrained by difficulties in matching the work undertaken by the current agricultural research system — a system developed under a centrally planned economy — with the needs of the new socialist market economy. The latter requires a more open, flexible research system, more closely linked to commercial customers and able to address the more complex interactive problems associated with the changing nature of today's agriculture.

Chinese comments

Following the 1995 Decision on Accelerating Scientific and Technological Progress, there have been a number of new trends in agricultural research, which, if continued, will address some of the comments raised in the Mission's report. The following are the most significant of these trends:

- The more frequent interaction of agricultural scientists with end-users of the technologies they produce and, in turn, the growing appreciation by farmers and farming enterprises of the need to incorporate S&T considerations into their farming and commercial activities; and

- The demand created by new rural enterprises for new technology from agricultural research, which is beginning to outstrip the supply, with the result that local groups of farmers and TVEs are organizing their own R&D groups to supply technologies to existing and new enterprises and production systems.

In the planning for the next 15 years (to 2010), there is emphasis on

- Strengthening the technology-evaluation and extension roles of research institutes;

- Focusing research efforts on key agricultural problems selected on the basis of regional priority; and

- Strengthening the research capacity of agricultural research institutes and increasing financial support for their activities.

Topic 7 — Basic Research

Mission comments

The Mission drew on its main report to identify five aspects of basic research it considered particularly important for Chinese policy:

- The definition of basic research;
- Resource allocations for basic research;
- The relationship between research intensity and doctoral training;
- Management challenges in the new national research centres; and
- The definition of the role and program scope of CAS.

Each aspect was elaborated in the presentation, and the policy challenges were identified.

Chinese comments

Two main issues were addressed. These were an analysis of factors influencing Chinese policy on basic research and a discussion of the financial support provided to basic research in China. For China, basic research has several functions:

- Providing a motive force and the knowledge base needed for social and economic development;

- Exploring new areas of knowledge and promoting enhancement of knowledge;

- Fostering the development of highly qualified personnel; and

- Being a significant source of national pride and prestige.

The key features of the financing of basic research are the following:

- There has been sustained and steady support via the budget of NNSF;

- Competition for funding has been used to avoid duplication of activities;

- Three levels of support are available (for general projects, key projects, and major projects, each of increasing scale);

- NNSF is intimately engaged in fostering the training of talented people, in addition to financing research; and

- Increasing levels of cooperation between enterprises and research institutions have been promoted.

Among the continuing issues of concern in policy for basic research are the following:

- How to promote more interdisciplinary activities and how to break away from too much reliance on activities in traditional disciplines;

- How to increase the number of talented people in S&T and how to make them more innovative; and

- How to cope with the huge numbers of applications for NNSF support (the success rate of applicants seeking grants today is about 16%, and NNSF annually receives more than 20 000 applications for review).

During the discussion that followed, it was indicated that ISPM–CAS has been doing internal studies of the work of CAS and found evidence of a continuing and important trend toward focus on strategic and applied research.

Topic 8 — High Technology

Mission comments

A significant result of the reforms of the last decade has been the increase in the contribution of high-technology industries to industrial growth — the Mission estimated that the contribution increased from 2% of GDP in

1985 to 10% in 1995, with the expectation of 15% in 2000. This has resulted from two sources — namely, indigenous technology being capitalized in China and imported technology being either adapted or implemented in joint ventures.

In the case of indigenous technology, there has been a significant change in research culture as a result of funding pressures on research institutes and universities. Many young entrepreneurial scientists have moved out to start up high-technology enterprises. There have been benefits to these scientists: increased salaries; freedom of operation; and freedom to pursue new developments. However, there is some concern that they lack training in management and finance, and the suggestion was made that the failure rate of businesses could be reduced by suitable training in universities and research institutes.

As a result of the open-door policy, there has been a massive importation of technology (much of it high technology) to rapidly increase productive capacity. There appears to be a perception that imported technology is always superior to indigenous technology, but this is not always the case. An indication of technological dependency is the ratio of payments for imported technology to R&D expenditure — the lower the ratio, the lower the dependency. For Japan it is 6%; for Korea, around 20%. For China, the figure is clearly much higher (data seem to be sketchy, but from what we heard in discussions, the ratio could be as high as 200%!). China must aim to reduce this dependency as a matter of policy; otherwise, the inevitable drying up of technology transfer will cause a major shock. This can be achieved by the projected buildup in the GERD–GDP ratio to 1.5%, coupled with strong encouragement for the exploitation of indigenous technology.

Other topics related to high-technology industries raised by the Mission included

- The problem of duplication in the numerous high-technology programs, such as the 863 Program, the National Key Laboratory Program, and the Climbing Program, the ERCs, CAS, and the universities;

- The nature of the interaction between universities, industry, and government in China; and

- The proliferation of development zones in China.

Chinese comments

The Chinese comments were limited to a description of the 863 Program as a leading policy instrument to promote high-technology industry. The objectives of the 863 Program are

— To aim for the highest levels of world high-technology development and to narrow the gap between China's attainments and world levels;

— To train a contingent of highly creative high-technology talents;

— To commercialize and put into production R&D results in order to promote the transformation of traditional industry, lay the foundation for new high-technology industries, and contribute to the national economy and state security through to the next century; and

— To promote nationwide progress in high-technology development, to create the conditions for building fairly developed high-technology industries after 2000, and to prepare for steady and sustainable growth of the national economy toward significantly higher standards.

To work toward these goals, the 863 Program has adopted a series of policy guidelines:

— Leadership of projects is to be unified, and resources are to be concentrated;

— Activities are to be effectively coordinated, and links with application are to be improved;

— Project management and decisions on granting support are to be the responsibility of experts and specialists;

— Real responsibilities are to be given to young and middle-aged specialists;

— Specific products are to be targeted for production; and

— International linkages are to be promoted.

At the conclusion of the 2 days of formal meetings in Beijing, it was agreed that the process of conducting the review, the contents of the Mission report, and the wide-ranging discussions the report had stimulated had all made a useful contribution to thinking in China about the way ahead for S&T policy reform.

Discussions in Shenyang, Xi'an, and Shanghai

Following the Beijing seminar, the Mission team divided into three groups, returning to three of the cities visited in November 1995. Meetings were organized in each of the cities by the local STC and in all cases included STC representatives, as well as individuals from business enterprises and institutions of higher education. The discussions in these cities addressed the substantive issues raised in the Mission's report, as well as providing a more detailed critique of the report itself.

There was a general consensus that the report had provided a useful overview of the S&T policy reforms in China and had helped some of the provincial officials put their own work in a wider national and international perspective. Many of the participants in these regional meetings agreed that the problem of duplication of S&T activities, along with the lack of coordination, was a major one.

There were some interesting differences between the comments made in the regions and those made at the discussions in Beijing. For example, the main criticism of the report in the regions was that the Mission had correctly identified weaknesses in the Chinese S&T system but had not recommended solutions. This was seen as a serious shortcoming. The Mission's response was that the objective of the review was to help diagnose strengths and weaknesses but not to recommend solutions, which would have been presumptuous for a foreign team that spent only 3 weeks in China. The team had, however, suggested in its report a number of areas in which it would be useful for Chinese policymakers to study foreign experiences.

The other major difference between the discussions in Beijing and those in the regions was that in the latter there was a great deal of interest shown in the concept of an NSI. It was felt that this might provide a framework for thinking about the next phase of reform policies. The debate on this topic was more enthusiastic in the regions. Subsequently, though, SSTC in Beijing has given this topic a high priority for follow-up activity.

The discussions in the regions also produced more examples that tended to confirm rather than refute the Mission's conclusions.

Postscript

On a visit to Beijing in September 1996, Mr Keith Bezanson, President of IDRC, presented a draft of the Mission report to Madam Zhu Li-lan, Executive Vice Minister of SSTC. On that occasion, Madam Zhu commended the Mission for its work and said that its findings had already influenced government policy. Two topics were identified as warranting further investigation. These were the NSI and the experiences of other countries in developing national strategies for international collaboration.

Appendix 1

TEAM OF EXPERTS

J.R. McWilliam
129 Mugga Way
Red Hill, Canberra
ACT, Australia 2601
Phone/fax: (61 75) 444 0755

Jim Mullin
Mullin Consulting Ltd
44 Grierson Lane
Kanata, ON, Canada K2W 2A6
Phone: (613) 839 1990
Fax: (613) 839 1707
E-mail: jmullin@compmore.net

Geoffrey Oldham
The Clock House
Barcombe Place
Barcombe, N. Lewes, BN8 5DL
Sussex, UK
Phone: (44 1273) 400 975
Fax: (44 1273) 401 517
E-mail: g.oldham@btinternet.com

Peter Suttmeier
Department of Political Science
University of Oregon
Eugene, OR 97401, USA
Phone: (541) 346 2856
Fax: (541) 346 4860
E-mail: petesutt@oregon.uoregon.edu

Greg Tegart
University of Canberra
PO Box 1, Belconnen
ACT, 2616 Australia
Phone: (61 6) 201 5230
Fax: (61 6) 201 5370
E-mail: gregt@ise.canberra.edu.au

Anthony Tsou
60 Jalan SS 22/35
Damansara Jaya
47500 Petaling Jaya
Malaysia
Phone: (60 3) 716 5936
Fax: (60 3) 713 5934

Anne V. Whyte
Mestor Associates
751 Hamilton Road
Russell, ON, Canada K4R 1E5
Phone: (613) 445 1305
Fax: (613) 445 1302
E-mail: mestor@sympatico.ca

Appendix 2

ACRONYMS AND ABBREVIATIONS

CAAS	Chinese Academy of Agricultural Sciences
CAS	Chinese Academy of Sciences
CAST	Chinese Association for Science and Technology
CIMS	computer-integrated manufacturing system
CPC	Communist Party of China
EPA	environmental protection agency
ERC	engineering research centre
GDP	gross domestic product
GERD	gross expenditure on research and development
GNP	gross national product
HRD	human-resources development
IDRC	International Development Research Centre
IPR	intellectual property rights
ISPM	Institute of Science Policy and Management [CAS]
NNSF	National Natural Science Foundation
NRCSTD	National Research Centre for Science and Technology for Development
NSI	national system of innovation
NTE	new technology enterprise
OECD	Organisation for Economic Co-operation and Development

R&D	research and development
S&T	science and technology
SCRES	State Commission for Restructuring the Economic System
SEC	State Economic Commission
SEdC	State Education Commission
SETC	State Economic and Trade Commission
SOE	state-owned enterprise
SPC	State Planning Commission
SRCAP	Shanghai Research Centre for Applied Physics
SSTC	State Science and Technology Commission
STC	Science and Technology Commission
TVE	township and village enterprise
WIPO	World International Property Organization
WTO	World Trade Organization

BIBLIOGRAPHY

Broadman, H.G. 1995. Meeting the challenges of Chinese enterprise reform. World Bank, Washington, DC, USA. World Bank Discussion Paper 283.

China Daily. 1995. Put money into fittest operations. China Daily, 4 July, p. 4.

Fan, S.; Pardy, P.G. 1992. Agricultural research in China: its institutional development and impact. International Service for Agricultural Research, The Hague, Netherlands. 96 pp.

Fang Xin n.d. Analysis and evaluation on the management system of science and technology of China. Chinese Academy of Sciences, Beijing, China. CAS Evaluation on the Strategy of Science and Technology Development of the Chinese Government series, Report 3.

Gibbons, M.; Limoges, C.; Schwartzman, S.; Scott, P.; Trow, M. 1994. The new production of knowledge: the dynamics of science and research in contemporary societies. Sage, London, UK.

Government of China. 1996. Evaluation of the key research and development program in the Eighth Five-Year Plan (1990–95). [In Chinese]

IOSC (Information Office of the State Council, China). 1995. China: arms control and disarmament. IOSC, Beijing, China.

National Affairs. 1995. The CPC Central Committee and State Council Decision on Accelerating Scientific and Technological Progress: Section VI. National Affairs, FBIS-CHI-109, 7 June, p. 26.

Organisation for Economic Co-operation and Development (OECD). 1991. Technology and productivity: the challenges of economic policy. OECD, Paris, France.

——— 1992. Technology and the economy: the key relationships. OECD, Paris, France.

——— 1994. National systems of innovation: general conceptual framework. OECD, Paris, France. DSTI/STP/TIP(94)4. p. 3.

SPC and SSTC (State Planning Commission and State Science and Technology Commission). 1994. China's Agenda 21. White paper on China's population, environment and development in the 21st century. China Environment Science Press, Beijing, China.

SSTC (State Science and Technology Commission). 1986. Decision of the Central Committee of CPC on the reform of the science and technology management system. *In* Guide to China's science and technology, 1986. White paper on science and technology No. 1. China Academic Publishers, Beijing, China. pp. 395–404.

——— 1994. International Symposium on China's Spark Program. SSTC, Beijing, China.

State Council. 1995. Regulations on transforming the management mechanisms of state-owned industrial enterprises. *In* Chen Junsheng, ed., China's reform policy canon. Hongqi Press, Beijing, China.

STC–CPPCC (Science and Technology Commission of the Chinese People's Political Consultative Conference). 1994. The problems and proposals concerning technological progress in large and medium-sized state-owned enterprises. National Research Centre for Science and Technology for Development, Beijing, China. Forum on Science and Technology in China, Vol. 1, No. 1, June 1994.

Tegart, G. 1995. The Co-operative Research Centres Program. University of Canberra, Canberra, ACT, Australia. Nexus Report.

Yang, L. 1994. Present situation and future development of science and technology in China. National Research Centre for Science and Technology for Development, Beijing, China. Forum on Science and Technology in China, Vol. 1, No. 1, June 1994.

Xu Zhaoxiang. 1995. Policy and institutional priorities for industrial technology development: China. National Research Centre for Science and Technology for Development, Beijing, China. Forum on Science and Technology in China, Vol. 1, April 1995, p. 83.

World Bank. 1995. Staff appraisal report: China — Technology Development Project. World Bank, Washington, DC, USA. Report 2814-CHA.

About the Organization

The International Development Research Centre (IDRC) is committed to building a sustainable and equitable world. IDRC funds developing-world researchers, thus enabling the people of the South to find their own solutions to their own problems. IDRC also maintains information networks and forges linkages that allow Canadians and their developing-world partners to benefit equally from a global sharing of knowledge. Through its actions, IDRC is helping others to help themselves.

About the Publisher

IDRC Books publishes research results and scholarly studies on global and regional issues related to sustainable and equitable development. As a specialist in development literature, IDRC Books contributes to the body of knowledge on these issues to further the cause of global understanding and equity. IDRC publications are sold through its head office in Ottawa, Canada, as well as by IDRC's agents and distributors around the world.